BELLWETHER HISTORIES

EDITED BY SUSAN NANCE
AND JENNIFER MARKS

Bellwether Histories

Animals, Humans, & US Environments in Crisis

UNIVERSITY OF WASHINGTON PRESS SEATTLE

Bellwether Histories was made possible in part by a grant from the Samuel and Althea Stroum Endowed Book Fund.

The open access publication of chapter 4, "Animal Photography and the 'Elk Problem' in Modern Wyoming," was funded by the Netherlands Organization for Scientific Research (NWO) VICI project Moving Animals (VI.C.181.010). DOI: 10.6069/9780295751436.04

Design by Mindy Basinger Hill
Composed in Garamond Premier Pro

UNIVERSITY OF WASHINGTON PRESS *uwapress.uw.edu*

LIBRARY OF CONGRESS CATALOGING-IN-PUBLICATION DATA
Names: Nance, Susan, editor. | Marks, Jennifer (Historian), editor.
Title: Bellwether histories : animals, humans, and US environments in crisis / edited by Susan Nance and Jennifer Marks.
Description: Seattle : University of Washington Press, [2023] | Includes bibliographical references and index.
Identifiers: LCCN 2023006022 | ISBN 9780295751412 (hardcover) | ISBN 9780295751429 (paperback) | ISBN 9780295751436 (ebook)
Subjects: LCSH: Animals and civilization—United States. | Human-animal relationships—United States—History—19th century. | Human-animal relationships—United States—History—20th century. | Human ecology—United States—History—19th century. | Human ecology—United States—History—20th century.
Classification: LCC QL85.B435 2023 | DDC 590—dc23/eng/20230302
LC record available at https://lccn.loc.gov/2023006022

♾ This paper meets the requirements of ANSI/NISO Z39.48-1992 (Permanence of Paper).

FOR ACE AND THE OLD LURCHER, HUGGY BUG

CONTENTS

PREFACE

This project began during a global pandemic and was inspired by the intersection of two books. First inspiration came from David Quammen's comprehensive summary of the current state of global ecology in the final pages of *Spillover: Animal Infections and the Next Human Pandemic*, in which he explains how human population growth, wild animal habitat destruction, and human settlement in those spaces guarantee that pandemics are here to stay. Next, we thought of Alanna Mitchell's *Seasick: Ocean Change and the Extinction of Life on Earth* and her account of the "serial extinctions" phenomenon studied by Heike Lotze and Richard Hoffmann. In medieval Europe, when people exploited preferred fishes to extinction, rather than cutting back on family size or ignoring a religious edict to eat fish, for instance, they invented new technologies and economic, legal, and food systems to reach fish populations farther afield. Putting together the work of Quammen and Mitchell helped us realize that our current moment of climate and environmental crisis has been many centuries in the making. Human overextension and its ecological consequences are nothing new. For how long have we imagined that we could harm animals without harming ourselves? The history of people enacting serial extinctions and other damage to animals and their spaces—as well as belief systems and decision-making frameworks that facilitated it—requires more study. This book thus focuses on the linkage between the ideologies of animal disposability and human exceptionalism and offers American stories in this larger history.

We would like to thank our editor, Andrew Berzanskis, and all the production staff at the University of Washington Press for supporting this project and offering crucial advice at every turn. A hearty thanks also to the anonymous reviewers for their likewise constructive critiques and guidance. Their insights helped make the book more cohesive and pushed us to articulate some of its more abstract concepts. Thanks to Dr. Katrina Merkies of the University of Guelph and Kayla Johnson of the Donkey Sanctuary of Canada for their help interpreting figure 3 in the book's introduction. With thanks, Vanessa Bateman's chapter, "Animal Photography and the 'Elk Problem' in Modern

Wyoming," is part of the Moving Animals: A History of Science, Media and Policy in the Twentieth Century Project (VI.C.181.010), financed by the Dutch Research Council (NWO). Lastly, as always, we are very grateful to the colleagues, friends, and family—both nonhuman and human—who helped us navigate this work in complicated times.

BELLWETHER HISTORIES

FIG. 1.1. *A Lonely Job: Waiting All Alone in the Dark for a Job to Come Through*. Lewis Wickes Hine, photographer, January 1911. National Child Labor Committee Collection, Library of Congress, Prints and Photographs Division.

SUSAN NANCE AND JENNIFER MARKS

introduction The Mule in the Coal Mine

Between 1908 and 1924, the progressive social scientist and photographer Lewis Hine traveled to coal country to document child laborers in the places where they worked, producing a portfolio of deeply empathetic, politically explosive photographs.[1] Hine interviewed his subjects directly, inquiring about their lives and experiences to discover that few attended school. Many worked illegally, since parents and employers evaded labor laws by falsifying young workers' ages.[2] In many places, Hine encountered children enthusiastic about their jobs and earning power. Sent by their parents to work dangerous or unhealthy jobs, these children were supervised by adult mine employees who were likewise mired in poverty and exploitative labor systems. In such a context, few adults believed that child miners had a right to any other life.[3]

On one visit to the anthracite coal mine in Pittston City, Pennsylvania, without formal permission from management, Hine talked his way down into the mine. There he encountered thirteen-year-old Willie, who worked as a door boy, also known as a "trapper" or "nipper." Willie's boring but crucial job was to wait in the dark to open and close a wooden door. Mine ventilation depended upon such systems to control flows of methane and facilitate the ingress of fresh air. Hine captured Willie in profile, sitting in the darkened space, normally illuminated only by an open-flame oil torch attached to the bill of his hat (fig. I.1). A single coat hanging above his head, Willie faces the wall, isolated and dwarfed by the great wooden door and expanse of cart track. Hine watched as Willie waited in the dark silence until hoofbeats and the voices of teenage mule drivers were heard from the other side.

Children entered coal work around age ten as "breaker boys," picking out by hand bits of slate and other rock from conveyor belts of raw coal hauled above ground to a sorting facility. In adolescence, they graduated to "nippers" underground. Then, as teenagers, they often won promotion to work as drivers, managing mules and ore carts in the mines (fig. I.2).[4] These boys were materially reshaped by their labors, suffering chronic lung and eye problems as well as stunted development that gave them characteristically rounded shoulders and small stature.[5]

Mules found themselves sucked into the world of anthracite coal because mining companies preferred them over other equines for pulling ore carts. Stronger than burros, mules were also hardier and, some said, wiser than horses to the dangers of the mines. There was a substantial industry in Kentucky and other midwestern states breeding mules for sale to mining companies because mine mules often died or became too injured to work while very young, could not reproduce, and required frequent replacement.[6] The industry took only black and dark bay mules (not gray or sorrel, whose appearance in the dark was believed to frighten other mules underground), and favored big-boned mules of shorter than average height to navigate the tunnels with heavy loads and fewer head injuries.[7]

Just as it did with the children who worked in those spaces, anthracite coal extraction produced a particular kind of mule in numbers that would not have overwise existed—a mule shaped by his or her context, with coal dust in sinuses and lungs, eyes unaccustomed to natural light, and chronic soft tissue injuries. Mine mules were also shaped behaviorally: they were especially known for kicking with "fiendish delight" equipment, stable walls, their drivers, and the other men and boys who worked in the dark with them.[8] Said one advocate for the mules after listing the conditions in the mines, the overwork, and frequent abuse from teenage drivers, "There is small wonder that they become vicious, and kick and bite at every opportunity."[9] Did mine mules kick just to aggravate, as some miners said, or because they could not properly see what was around them, out of fear or frustration, or for some other reason?

Underground, new ecologies developed as a web of species, human children among them, some of which were united by their vulnerability to exploitation by human adults while others were there seemingly by choice. Miner folklore remembered colonies of mice and young miners' efforts to befriend and feed them, as well as robust populations of rats. Both those rodent communities unknowingly served the proverbial "canary in the coal mine" function at times and were known to scatter in the minutes before a cave-in or to collapse just before a passageway flooded with toxic gas.

Thus, throughout Appalachia, recent immigrants from Europe, people of color, and rural folk with little formal education and chronic experiences of

FIG. 1.2. *Holding the Door Open while a Trip Goes Through*. Lewis Wickes Hine, photographer, January 1911. National Child Labor Committee Collection, Library of Congress, Prints and Photographs Division.

poverty tunneled into mountains whose Susquehannock inhabitants had been killed or scattered by warfare and fraud over a century earlier. They created the spaces where these ecologies of the vulnerable developed—hills made vulnerable by their coal seams, children made vulnerable to temptation to mining work by poverty, and mules born vulnerable because they could not act collectively to strike or escape.

Adult miners could quit or go on strike. Child miners were notorious for fooling around, tampering with equipment, and otherwise resisting by slowing the pace of work. In their own way, mine mules also exposed the problematic nature of the industry. When first brought to the mine entrance at about two years old, mules often desperately resisted being cast with legs tied, then hauled into the mine in a coal car or lowered down in the elevator cage, braying "in terror" as they sank into the darkness.[10] Once at work, mules were handled by

FIG. I.3. An inconclusive photograph: underage miners with apparently underweight mules showing overgrown hooves, one with swollen fetlock, greasy coat, and strained facial expression. *Drivers Going Home, 5:30 P.M. from Mine Macdonald, W. Va.* Lewis Wickes Hine, photographer, October 1908. National Child Labor Committee Collection, Library of Congress, Prints and Photographs Division.

teenage drivers with little training, who were expected to break in new mules and get them to navigate sloped tracks with heavy loads. Mules sometimes broke legs when run over by coal cars to which they were harnessed or were otherwise fatally injured by their wagons and cargo. Many worked lame, with sore joints from overwork on uneven surfaces and hoof injuries from inexpert farriers, filthy stables, or kicks to stall walls when irritated and frightened by the rats who ran across the mules' backs. From improperly fitted tack, some displayed festering sores that drained pus and blood down their legs and were said to smell so terrible they made cart boys vomit (fig. I.3).[11]

Proindustry voices plainly described these conditions as inevitable aspects of mining work, as though they could not imagine a different way of doing things. In those days, coal was central to the productivity and day-to-day comfort of the nation. It fueled the railways and heated the majority of private and public buildings. American urbanization, industrialization, and economic growth depended upon coal, while mining it relied upon inequalities among Americans and power imbalances between species that kept coal prices low. Animal advocates and other political progressives explained mine conditions as a function of capitalism. "The coal companies do not care a tinker's cuss about cruelty to dumb [voiceless] animals," said one advocate. "A cold blooded coal corporation that practices social crimes on human beings, is not expected to entertain any high notion of the justice of things, even in the brute creation."[12] In this for-profit work space, management cut corners, staff were minimally trained in animal management and were themselves vulnerable to exploitation, while child and animal advocacy organizations and government inspectors had little access.

Human beings are curious, capable, and fond of inventing things and going new places. As we create unforeseen consequences, we adapt socially and psychologically to perceive that new situation as normal. Yet there are always dissenting voices, the people who ask, "What are we doing here?" On its editorial page, one Pennsylvania newspaper lamented the comprehensively broken-down state of the animals and people in coal mining, as well as the impression that miners and management seemed not to question the situation:

> We have seen mine mules hobble from their stalls in the morning on three legs, holding one stiff and sore member dog-fashion, until forced to put it on the ties when driven to labor. We have seen a mine foreman run after a driver boy and his mule, as they came out of the barn in the morning, and ask what had happened to one of these working cripples, but when told, answered "O, that's nothin.' Old Doll always carries one leg in the air unless she has to pull," [and] let boy and mule pass on to their work.
>
> We have seen mine mules working from early morning until evening without a bite to eat or a drop to drink between times. We have seen mine mules reach ravenously into the dirty ditches to lap with their tongue the black muck, or gulp a mouthful of sulphur water. We have seen driver boys urge them on past the ditches in their hurry to deliver their quota of cars in the time allotted to them. We have seen mine mules fall and die in their harness from sheer exhaustion on their way to the barn at night.[13]

Humane organizations advocated for mine mules, but only to argue for better welfare—not to abolish their use outright, nor to abandon coal-powered comfort and its economic opportunities.[14] The historical record is largely silent on the few observers who undoubtedly believed it was unethical to purposely breed animals known to prosper with sunlight, clean air, and fresh forage, then confine them for years in a dark, coal-dusted tunnel. The notion of being responsible for the care, needs, and long-term viability of that which—or those who—fed one's family was not the rule in coal country or in most industrial settings. In part, this book is about that history of animal disposability and its ideological twin in US history, human exceptionalism.

The working conditions of the children in the mines horrified the American public, but the unnatural existence of mules there less so. These mules were representative of larger patterns in animal history; their miserable state was a dead-canary warning about which the Americans at hand mostly shrugged. Their misery was one among many signals that people ignored or failed to recognize—indications that many areas of the nation were to become mismanaged, denuded, and increasingly toxic over time. Despite Hine's photographs and other alarming reports, only electrification of the mines and relatively inexpensive and less (human) labor-intensive compressed-air locomotion would relieve mules of their obligation to the coal industry.[15] At the time, the number of mules employed in the industry was actually increasing—up to about 15,000 per year in the mid-1890s, although each mule carried less coal overall.[16] Soon, retrieving ore from the increasingly inaccessible, slim seams that remained required mechanical means beyond a mule's capability, so technology and the profit motive won the day.[17]

The plight of mine mules mostly bred self-interested nostalgia. A decade before Hine visited Pittston City, people spoke of the "passing of the long-suffering pit mule" with pity but no sense that lessons might be learned about how to reform human use of other species.[18] Instead, people spoke with wonder of the long hours in the dark and mules' ability to predict and respond to the routines of mine work in terrible conditions. Or they told sentimental stories of boys who sought comfort by huddling with their mules during explosions or cave-ins and waited to die together inside an injured mountain.[19]

Over a century later, the coal industry in the United States is winding down. In many places, coal extraction now takes the form of mountaintop removal, a mechanized and cost-efficient technique that destroys nonhuman and human habitat. Today, we recognize how the extinctions and ecological breakdowns we created in such ways have guaranteed that deadly zoonotic pandemics and global warming will harass us far into the future. Nonhumans have been forced along this path with us. Reflecting on this shared journey gives us the opportunity to examine the parallel and intertwined histories of mountains, human children, and equines, and to examine what Juno Salazar Parreñas has called our "mutual but unequal vulnerability in an era of annihilation." To "decolonize" our scholarship, she says, we must document what people have asked of nonhumans in the historical processes that unmade our environment in ways that animals often did not understand and could not evade. Only then can we come to terms with what will follow without imposing upon other species the unreasonable responsibility of somehow turning back the clock.[20]

The abuses of Pittston City's turn-of-the-century coal economy is just one of countless historical crises that reveal how, when problems emerged or disaster struck as colonization, capitalism, and environmental change combined in ways that were unexpected (to some), people did not substantially change course but sought cheap or convenient half-remedies, or did nothing at all. This book interrogates animal bellwethers who were ignored or discounted in American history in order to better understand the past without becoming nostalgic, uniformly demonizing human actors, or proposing a rigid solution for the future.[21] Together we document the historicity of human exceptionalism—specifically, the anthropocentric myth that people could harm animals and their spaces without harming themselves.

This book begins by mapping out the broad history of animals with a globalized, roughly American history of species wherein we include humans in the group "species." Any multicentury narrative of animal history in the United States arguably should begin with the colonization of North America. This process constituted a systematic attempt by individuals, businesses, and governments to wrest control of the environment from Indigenous and Hispanic peoples, animals, plants, water, and other historical actors. Thereafter, decade by decade, century by century, animals were prevented from exercising their autonomy. For many, that struggle for autonomy meant confronting habitat insecurity or outright destruction, population declines, and disease. For others, restriction of autonomy came with explosive population growth combined with economic exploitation and intensive confinement.

This book documents animals confronting the Anthropocene, that combination of colonization, capitalism, technological development, and human population growth that, as Donna Haraway puts it, "has changed the planet forever and for everyone."[22] As an analytical concept, *Anthropocene* is both useful and problematic in understanding what nonhumans have encountered over the last several centuries. The term rightly reminds us that humans have been a dominant, if not the primary, actor in producing the climate crisis and sixth extinction, as well as the many local and regional disruptions that circumscribed animals' autonomy. At the same time, *Anthropocene* privileges human agency and risks minimizing the very interdependence of living beings this book seeks to expose.[23] Still, people caused the historical conundrums and crises this book documents, and they often relied upon their belief in human exceptionalism to reconcile the contractions and dilemmas that ensued.[24] Americans debated what was happening, why, and what, if anything, to do. We are interested in understanding the decisions they made and the historical patterns those decisions created over time. Like the mine mules with their ability to cope in difficult conditions, the resilience of the animals documented in this book often obscured just how shortsighted and destructive human plans and actions were.

We contend that colonization is a crucial ingredient to explaining American animal history and how the myths of human exceptionalism and the practices of animal disposability came to dominate American life. The colonization process, its politics, ideologies (like Christian dominionism), capitalist economics, and violence inspired settlers of all kinds to systematically deny, ignore, or intentionally destroy ecologies and relations among nonhumans and between human and nonhuman species.[25]

American efforts at capitalist development and nation-building began with a colonial approach toward Turtle Island and its inhabitants. Like the many people on the continent who struggled against racialization and linked forms of environmental injustice, animals lacked environmental rights to a particular habitat, freedom from human contact or manipulation, and freedom from the impacts of human activity. Unlike historical people who understood the systems of power, terror, and theft against which they struggled, animals have never understood the causes of organized human activities and have suffered the consequences accordingly. For a few—like the mice and rats in Pennsylvania's coal mines—there were opportunities to capitalize on the changes human activities produced. Many diseases can be counted among those more fortunate opportunists.

Americans are part of an interspecies community that scholars neglected for much of the twentieth century by writing history as a record of human agency and exceptionalism.[26] More recent scholarship has embraced the interspecific relationality theorized and practiced by the many Indigenous peoples who approached other species with humility and a sense of responsibility for a shared fate. These works also draw on the ideas of Aldo Leopold and other early settler environmental thinkers who asked their contemporaries to respect other species' right to exist.[27]

The stories in this book complement that scholarship by examining some of the many cases where settlers discussed these same ideas or had similar realizations but did not change course. Some would walk away from an animal disaster vowing never to return—for instance, ranchers who left the cattle business after the Big Die-Up, discussed in chapter 3 by Susan Nance. Others sought remediation or band-aid solutions to remedy the symptoms of a problem that seemed sadly unavoidable, such as the Leek family, who, as Vanessa Bateman explains in chapter 4, fed hay to starving elk endangered by settlers. Others saw opportunity in ecological disruption, like the professional insect managers in Hawai'i whom Jessica Wang documents in chapter 5. Still others, like Pennsylvania's mule drivers, saw animals in crisis but perceived no crisis at all, so carried on with no change.

The authors in this collection spotlight the ties that bind human and animal fates and chart the patterns of unintended consequences that resulted when Americans denied those ties. We compare interspecies relationships across cultural, temporal, and political contexts. In doing so, we reveal the repeated errors and compromises people made and the warnings they ignored. Animals were speaking in their own ways all the while, but they were seldom heard.

Our work to understand these problems is complicated by historical sources created by people in highly anthropocentric contexts and archived in institutions that begin from an assumption of human preeminence. Mindful of these challenges, we address key methodological questions for animal history in our age:

1. How do we find the animal experience in human sources? And how can we identify anthropocentrism in how sources were created and what they appear to show now?

2. What factors bind together human and animal histories? Under what circumstances do we find evidence in sources of that "mutual but unequal vulnerability" between humans and others in an environment?
3. Where do human power structures—colonization, race, gender, or global imperialism—intersect with and affect the lives of nonhumans? How do inequalities among people reflect and shape interspecific relationships?

The chapters here are bookended by two globalized examples, specifically eighteenth-century debates over the transatlantic movement of livestock to colonized lands and, in the later twentieth century, the fate of city zoo captives in an American imperial setting in Iraq. These stories remind us that the United States has never been a closed system; it is located within larger transportation, trade, and information networks that began in the late fifteenth century and facilitated America's rise as a global superpower. Animals have always been uninterested in international political boundaries in any event.

The arc of animal history in the globalized United States is one of struggle to exercise autonomy in the face of human population growth and landscape revision. Chapter 1 drops us into the story when the United States was a fledgling republic within an imperial world system financed and facilitated by the circulation of capital and commodities, including people and livestock. Here, Joshua Kercsmar investigates the foundational links between antislavery and anti–animal cruelty advocacy, explaining how, in the context of the market revolution, racialization and animalization were parallel political concerns to many thinkers watching trade around the Atlantic. American national security was rooted in an economy of colonization and slavery. According to Kercsmar, these systems profited off the commodified circulation of bodies and a "mercantile logic [that] flattened differences between animals and enslaved people" and inspired some of the first of many reform movements.

On the ground on Turtle Island, animals possibly first detected the changes colonization would bring through human population declines as epidemics attacked Indigenous peoples. In the colonial and early national periods, especially in the east, many animal species confronted a rapid human repopulation as settlers made their way in from the coasts. This often meant more frequent encounters with humans, increased subsistence and market hunting, new species like horses and pigs, and widespread deforestation and draining of wetlands.

By the mid-nineteenth century, the United States had a number of growing urban centers reliant on captive equines who became vehicles for an epizootic in 1872. In chapter 2, Jennifer Marks retraces Chicago's responses to the highly contagious respiratory disease, which capitalized upon Chicagoans' overreliance on large populations of workhorses. Rigid infrastructure, the high cost of alternative energy, and personal investments in existing systems all contributed to the human inertia that made cities vulnerable to the precarious welfare of horses and mules. This energy crisis so profoundly disrupted city function that many Chicagoans saw horses as living beings for the first time. Yet, like the case of mine mules, the epizootic would not be the last time Americans' investment in a large-scale animal exploitation system would break down and cause reflection, only to be patched up when people could not implement, imagine, or take seriously an alternative.

Next, the book follows the rail lines of American westward expansion past Chicago to rural Montana and Wyoming, where settlers were systematically unmaking ecosystems to make way for railways, settlers, state power, and the growing industrial network of resource extraction, manufacturing, and consumption. Bolstered by military victory during the Civil War (although still struggling for control against a violent white-supremacist insurgency in the South), Washington and the US military waged war against Indigenous nations.

Here, in chapter 3, Susan Nance documents the Big Die-Up, a human-made animal welfare disaster that marked early efforts to produce a beef monoculture in Montana to serve meatpackers in Chicago and beyond. Absentee investors sought to profit from growing markets for beef by grazing behaviorally naive, imported cattle on lands recently confiscated from Indigenous peoples forced onto reservations. Blindsided by many weeks of frigid temperatures and whiteout conditions in 1886–87, those cattle perished from thirst or exposure by the hundreds of thousands. Although range cowboys, who often futilely struggled to keep those cattle alive, recounted these traumatic events, the industry would muddle along. Settlers faced with disaster, on balance, persisted with the colonial folklore that they would master the environment next year. That idea was crucial to Americans' work to turn the continent's various regional or micro-ecosystems into producers of commodities and sites of family-owned agricultural businesses during the nineteenth century.

At the same time, settlers waged extermination wars on wild predators and competitor species like bison and wild horses. They were encouraged by state and federal programs designed to grow the economy, entrench settler power, and link the West to national economic networks. Elsewhere, settlers disrupted local animal migrations, feeding, and reproduction simply by occupying vital lands. Along with market hunting, many species across the Midwest and West experienced radical population declines and for some, like bison and wolves, functional extinction.

In these crises there were opportunities for advocates and professionals to assert themselves as agents for animals and a particular vision of the future environment. In chapter 4, Vanessa Bateman points us south from Montana to the shrinking habitats of wild elk populations in turn-of-the-century Wyoming. There, elk struggled with starvation as homesteads and a nascent tourism industry crowded them out of critical feeding grounds. Local settlers seeking to support starving elk engaged not in the poetry or political tracts of the antislavery era discussed in chapter 1, but in modern means of persuasion through images and early animal refuge advocacy. Bateman explores how ranchers and guides employed photography and painting that, though well meaning, "represented wildlife and the landscape as independent from humanity—portrayed as unspoiled and idealized." Photography and "camera hunting" sought to affect viewers emotionally without directly confronting unsympathetic local residents in an ecosystem undergoing swift transformation. That work portrayed wild animals as beggars in an essentially human environment, where Americans would have the power to decide which species survived.

Just before the crisis of Wyoming elk, Americans had lived through a transformative decade. In 1890, the US government destroyed the last armed Indigenous resistance to state power, then emerged as a hemispheric power during the Spanish-American War. In the midst of the political chaos and jingoism of the period, many Americans failed to notice that their government had taken over Hawai'i. Insurgent Anglo settlers in the then-independent kingdom had quietly requested that the islands be annexed into the United States so that, among other reasons, local sugar exports could access protected American manufacturing and consumer markets. (Recall that Joshua Kercsmar's discussion of sugar production takes place in a similarly colonial economy, in Jamaica).

In chapter 5, Jessica Wang takes us to turn-of-the-century Hawai'i, where a class of professionals stepped forward to capitalize upon the unintended consequences of annexation. Wang analyzes a "complex assemblage of people and parasites" to explain trial-and-error attempts with species introductions there. In Hawai'i, nonnative horn flies and agricultural science professionals—both invasive "biological agents"—found opportunities in that colonized ecosystem, then struggled with one another for control of it. Wang's analysis pays particular attention to the ways animals reacted to attempted "management," often thwarting or altering scientists' and cattle ranchers' plans for the landscape. The story of the horn fly reminds us that a few animals capitalized on the radical remaking of landscapes and habitat disturbance brought by colonization. Like urban coyotes, suburban deer, and the bears who congregated around the dumps at national parks in the twentieth century, some animals found more freedom and autonomy in human-built spaces.

Turn-of-the-century Hawai'i was a context similar to those explained by the first four chapters of the book, one transatlantic, one urban, and two rural. Across the first decades of the twentieth century, settlers and the government continued systematically disassembling ecologies on lands only barely wrested from Indigenous peoples, while Americans jockeyed over who would benefit from what had been captured. Just after World War II, decades of this type of activity had made the United States a global superpower. Not coincidentally, superpower status also helped produce the Great Acceleration in anthropogenic environmental change. Americans had extraordinary access to natural resources, new technologies, consumer markets, manufacturing capacity, and a colossus of an economy. Paired with the anticommunist politics of the Cold War, patriotic American economists, political thinkers, business leaders, and consumers advocated for limitless economic and population growth while rejoicing in the newly abundant throwaway culture of postwar consumerism. All levels of government subsidized these trends by damming much of the state of Tennessee, passing the Eisenhower-era Federal Highway Aid Act, and diverting the Colorado River to promote development in the Southwest. Numerous other local, state, and federal acts pushed small-scale agriculture and remaining wild species out of many places where they still survived.

By then, Americans were most likely to relate to the animals at hand not as hunters or workers but as consumers. Dense clouds of passenger pigeons and the family mule were replaced by horse racing, consumer pet-keeping, children's literature and television shows about anthropomorphized animals, and visits to the city zoo. Human incursions into wild or semiwild spaces sparked confrontations with animals who had previously had only limited, sporadic exposure to people. Such conflicts had been increasing on the continent and around the world for several centuries. At home, Americans were exposed to the refugees of this activity by way of the exotic animal trade. The business of the global circulation of animals originally emerged from colonial livestock exports and the specimen collecting of premodern naturalists. Now, in the modern age, the trade and display of captured animals for privileged urban audiences was still a function of colonization, formal and informal.

In chapter 6, Andrea Ringer explains that context and how city zoos functioned as contradictory evidence of human restriction of animal autonomy, commodification of animal bodies, and denial of that reality—all at once. By the mid-twentieth century, zoos had begun to position themselves as champions for animals endangered in the wild. These institutions remained net consumers of animal bodies in stocking their displays, of course. Ringer explains that zoo staff spoke about conservation while investing ideologically in animal "spectacle markets"—the show business of global habitat destruction and animal capture. They accustomed the industry and the public to the idea that many wild and exotic animals rightfully belonged in an enclosure in the United States, where people could wisely take charge of their destiny. Ringer reveals how hippos were the cash cow at New York's Central Park Zoo between 1880 and 1975. Documenting decades of hippo breeding from two named individuals, Miss Murphy and her breeding partner, Caliph, the program was so profitable, she says, that it "funded nearly the entire zoo for decades . . . [by] the near-constant display of mother and infant, as well as selling and shipping unwanted older calves." Ringer's analysis reveals how women zoo workers and the gender politics of caring for and trafficking in captive-born animals determined the lived experiences of animal mothers and their "zoo babies," presaging and monetizing the coming sixth extinction.

In the later twentieth century, the continental and global circulation of commodified animal bodies materially connected wild animals like hippos to trends in American industrial animal agriculture and, in time, zoonotic pandemics.[28] Like zoos, large-scale animal agriculture operations were products of and contributors to the disturbed ecological relations that marked the Great Acceleration. The new government and university-subsidized concentrated animal feeding operations (CAFOs) were a mirror image of the zoo. There, technicians mass-produced captive animals who had short and generally miserable but economically productive lives spent in extreme forms of confinement evaded even by zoo animals. Like the mules, workhorses, and cattle discussed earlier, pigs experienced new bodily forms devised to serve not their own welfare but a steadily growing consumer demand for cheap meat.

In chapter 7, Mary Trachsel picks up this story of livestock monoculture, showing how it appeared as a precarious business on confiscated lands out west. The long history of Iowa's hog empire—its pigs and its humans—reminds us of the precolonial Indigenous peoples and diverse ecologies that were fully overtaken in the twentieth century by humans, pigs, soy, and corn. Like the crowded, epizootic-breeding stable owners of industrializing American cities a century earlier, Iowa's CAFO operators (themselves vulnerable to corporate agriculture entities and pork-price volatility) made animals vulnerable to disease through intensive confinement. Instead of perceiving a fork in the road, one branch of which led away from problematic low-cost pork, Iowans chose to sacrifice the state's biodiversity, air quality, and soil health. That denuded environment suffered alongside abused immigrant packers and the pigs, creating a reflexive system of dangerous work and environmental damage that many Iowans came to see as natural and desirable because it was economically productive. Trachsel asks us to consider the misery of Iowa's hidden hog population—locked away in high-security barns—and the corresponding environmental degradation and human health implications in the state. Like the mules sequestered underground, Iowa's hogs have lived a hidden industrial history from which consumers chose to turn away.

Finally, in chapter 8 we turn back to the global American context. John Kinder documents the tragedy at the Baghdad Zoo in US-occupied Iraq just after the turn of the twenty-first century. Here we see again that through line—beginning in chapter 1—that repeatedly, believing they could act upon animals and their spaces without personal consequence, Americans' activities might generate criticism but were only minimally reformed by contemporary animal advocacy. John Kinder's study of crisis at the Baghdad Zoo in 2003 and 2004 takes place in the context of the globalization of city zoos. Over the twentieth century, officials and prominent citizens in many countries came to believe that exotic and wild animal collections were necessary for the prestige of any national capital. Zoos appeared not as evidence of the legacy of human destruction of habitats and abduction of animals from the wild but as signs of the ostensible sophistication and wealth of the community or its leaders.

Employing newly discovered first-person accounts from the Baghdad Zoo, Kinder exposes how American troops in the zoo undermined official US policy in Iraq by enacting violence on animals trapped there. Like Iraqi leaders before them, American occupation officials publicly claimed to be engaged in a "rehabilitation" of the zoo and its captives as a symbolic gesture reflecting broader US policy toward the entire nation of Iraq. Kinder explains "the corporeality of Baghdad's zoo animals: their messy biology, their inconvenient livingness, their hunger and thirst and pain," to show how geopolitical conflict struggled to improve the experiences of animals caught between human imperialism and diplomatic publicity seeking. Again, animal advocates appeared. Local Iraqi facility staff clearly understood how the zoo's suffering animals were being victimized by US troops and exploited for US foreign and military goals. Yet, constrained by the longer connections between captive wild animal management and human imperialism of various kinds, they operated within an international animal captivity economy premised on "human-constructed solutions to human-produced environmental destruction," and thus resisted radical change or abolition.

Taken together, these cases tell a broad story of American animal history that is a record of declining nonhuman autonomy. Animals confronted anthropogenic environmental change and unwittingly rode the currents of human politics, economic ambition, and social change. Their past is punctuated by Americans steeped in the ideology of human exceptionalism. Believing they existed independent of the welfare and survival of animals, people constructed systems of resource use, relations with the land, or animal exploitation too complex to change quickly or cheaply. Each system or practice became intertwined with others as time passed. When an unexpected—or a much-predicted—crisis occurred, some learned the lessons of that crisis by witnessing the animals caught in the fray. Yet modern human mobility meant that there were always new people coming to seek their own fortunes and ready to recommit to problematic practices and beliefs grounded in human exceptionalism and the disposability of animal life.

These stories have shown us that most people were slow to learn. They made the same decisions over and over, decade by decade, generation by generation, until forced to finally undergo a paradigm shift, or radical change in thinking only when human health, prosperity, or politics was irretrievably broken or in collapse.

NOTES

1. Kate Sampsell-Willmann, *Lewis Hine as Social Critic* (Jackson: University Press of Mississippi, 2009), 10–11.

2. In many states, local laws barred children under age twelve or sixteen from paid employment.

3. Hugh D. Hindman, *Child Labor: An American History* (New York: Routledge, 2002), 90–95.

4. Adolescent and teenage workers accepted lower wages than adults and completed tasks in cramped spaces or benefiting from the dexterity of small hands. Child miners were especially prone to industrial accidents caused by inattention—fooling around on the job or working fatigued—like being crushed by coal cars, being snagged or pinched in machinery, plus the explosions and cave-ins that killed adult miners. Hindman, *Child Labor*, 99.

5. Hindman, *Child Labor*, 100–103.

6. Michael R. Sprowles, "Mine Mules: Their Use in Coal Mines in the United States," *Mining History Journal* 18 (2011): 26–27.

7. "Mine Mule's Burden," *Wilkes-Barre (PA) Record*, July 31, 1905.

8. "Festive Mine Mule," *Miners Journal*, October 25, 1898. See also Sprowles, "Mine Mules," 29.

9. "Mine Mule's Burden."

10. See, for instance, "Passing of the Long-Suffering Pit Mule," *Pittsburgh Daily Post*, November 21, 1897.

11. Thomas G. Andrews, *Killing for Coal: America's Deadliest Labor War* (Cambridge, MA: Harvard University Press, 2008), 130. See also "Lecture on Care of Abused Mine Mules," *Plain Speaker* (Hazleton, PA), January 24, 1905; "Cruelty to Dumb Animals," *Wilkes-Barre (PA) Times Leader*, October 18, 1912.

12. "Cruelty to Dumb Animals."

13. "Cruelty to Dumb Animals."

14. Anti-"cruelty" advocates sought to limit or eliminate animal suffering they deemed "unnecessary," meaning they believed it did not produce value for people that could be generated another way. Like most Americans in that era, anticruelty advocates argued that people had a right to use animals but should limit animal abuse since its presence coarsened society and desensitized people to violence and abuse in society more broadly. For instance, many animal advocates agreed that animal suffering in the name of entertainment was usually "cruelty," but animal suffering that supported food or energy production was not. Susan J. Pearson, *Rights of the Defenseless* (Chicago: University of Chicago Press, 2011), 80.

15. "Festive Mine Mule Has Lost His Grip," *Miners Journal*, October 25, 1898.

16. "Festive Mine Mule Has Lost His Grip"; "The Mine Mule," *Carbondale (PA) Leader*, December 2, 1898.

17. As it goes with most extractive industries, people first exploited the most accessible sources, then had to go to more difficult lengths as reaching what remained became more time consuming, technically complicated, and expensive. In that decade, the most accessible coal beds were being run out, and seams were thinner (another reason to employ child laborers who could wiggle into small spaces) and farther from the surface. "The Mine Mule," *Carbondale (PA) Leader*, December 2, 1898.

18. "The miners themselves will be sorry to see the mule go, because there is a sort of companionship in his presence, but for his own sake they should be glad that the faithful old mine mule is to become a thing of the past and his hopeless round of toil end through the advent of mechanism, which will do the work cheaper and better." "Passing of the Long-Suffering Pit Mule," *Pittsburgh Daily Post*, November 21, 1897.

19. See, for instance, Susan Campbell Bartoletti, *Growing Up in Coal Country* (Boston: Houghton Mifflin, 1996), 37–39.

20. For instance, Parreñas points to forced mating between surviving members of critically endangered species that takes place at wildlife refuges as representative of misdirected attempts to undo the damage human development has caused in ways that only impose new indignities and anthropocentric control on those survivors. Juno Salazar Parreñas, *Decolonizing Extinction: The Work of Care in Orangutan Rehabilitation* (Durham, NC: Duke University Press, 2018), 66; see also 3. See also Ben A. Minteer, *The Fall of the Wild: Extinction, De-extinction, and the Ethics of Conservation* (New York: Columbia University Press, 2019), 127–28; Anna Tsing, *The Mushroom at the End of the World: On the Possibility of Life in Capitalist Ruins* (Princeton, NJ: Princeton University Press, 2015), 3–5.

21. Parreñas, *Decolonizing Extinction*, 8–9; see also Donna J. Haraway, *Staying with the Trouble: Making Kin in the Chthulucene* (Durham: Duke University Press, 2016), 38–39; Dolly Jørgensen, *Recovering Lost Species in the Modern Age: Histories of Longing and Belonging* (Cambridge, MA: MIT Press, 2019), 5–6.

22. Donna Haraway, "Anthropocene, Capitalocene, Chthulucene: Staying with the Trouble," presented at Arts of Living on a Damaged Planet conference, AURA: Aarhus University Research on the Anthropocene, May 9, 2014, https: //vimeo .com/97663518 and https: //anthropocene.au.dk//conferences/arts-of-living-on-a -damaged-planet-may-2014/; Haraway, *Staying with the Trouble*, 6.

23. Haraway, *Staying with the Trouble*, 33, 38–39, 44–46; Minteer, *Fall of the*

Wild, 117; Timothy Sweet, *Extinction and the Human: Four American Encounters* (Philadelphia: University of Pennsylvania Press, 2021), 6–7.

24. Sweet, *Extinction and the Human*, 2.

25. Dina Gilio-Whitaker, *As Long as Grass Grows: The Indigenous Fight for Environmental Justice, from Colonization to Standing Rock* (New York: Beacon Press, 2019), 12–13.

26. Haraway, *Staying with the Trouble*, 60, 64.

27. This is a growing literature. See, for instance, Gilio-Whitaker, *As Long as Grass Grows*; Margaret Robinson, "Animal Personhood in Mi'kmaq Perspective," *Societies* 4, no. 4 (December 2014): 672–88; Sweet, *Extinction and the Human*, 6–8, 132–34.

28. David Quammen, *Spillover: Animal Infections and the Next Human Pandemic* (New York: W. W. Norton, 2012), 515–16.

JOSHUA ABRAM KERCSMAR

one Interspecies Anticapitalism in English and American Humanitarian Writings, ca. 1800–1850

As a physician in Durham, England, Henry Evans Holder III spent much of his time trying to save lives, but as owner of Joe's River, a 276-acre Barbadian sugar plantation, he was also in the business of quantifying them. In October 1808, he found himself occupied with a difficult calculation, for which he sought advice from the Barbados Society for the Improvement of Plantership: How much cane could he grow the following year with "75 head of cattle, 15 asses, [and] 3 horses," if "to effect this [work] there are 143 negroes, 60 of whom are labourers"? Should he plant less cane to compensate for "46 negroes being sold off" unexpectedly, including "18 . . . workers," or should he plan for the current amount of cane and buy more enslaved people?[1] Such questions, repeated in various forms by other planters throughout the society's minutes, entwined the fates of animals and enslaved people by rendering them as quantifiable units of production.

But the logic that dictated Holder's interspecies accounting had less to do with some a priori notion of enslaved peoples' equivalence to animals via Aristotelian or other lines of thought, as scholars have at times emphasized,[2] and more to do with his knowledge that in order to profit from his enterprise, he needed to balance the correct numbers of donkeys, horses, cattle, field workers, and other categories of labor against the market conditions in a given year (or quarter). The need to abstract human and animal life, to render it legible on a balance sheet, only grew as increasing numbers of planters moved back to Britain from the late eighteenth century onward, leaving their plantations in the charge of overseers and attorneys.[3] Holder also fit this pattern. Though he lived in Barbados in 1808, he later moved to England, where he carried on his medical practice at Durham Infirmary and managed Joe's River at a great distance from his human and animal chattels. He sold these chattels along with the entire plantation toward the end of his life, transforming them, as his will attested, from fixed assets into "money and other securities in Barbados and elsewhere."[4] In this way, Holder proved in death what he had long known: his animals and enslaved people were private and saleable property, had a value dictated by the market, and could, at his bidding, generate wealth through financial instruments like securitized debt, which had evolved through the early modern period.[5] They were living capital.

From the standpoint of planters, the value of animals and enslaved people lay in their ability to produce the tobacco, rice, sugar, and cotton that fueled growing transatlantic markets. In Barbados, sugar was the prime cash crop. From the 1650s, when Barbadian planters transitioned from tobacco to sugar, until the Slavery Abolition Act in 1834, enslaved people did the grueling work of planting, harvesting, and processing cane. Mules, oxen, and horses produced manure for the fields, hauled carts, and turned the cylinders of the mills that squeezed juice out of the cane. The brutality suffered by humans and animals is evidenced by their short lives. Enslaved people died so quickly that planters found it necessary to import new ones each year rather than count on self-sustaining populations. And plantation mules lived only six to eight years on average, while oxen survived only four to six, a fraction of their normal lifespans.[6] If the brutality of plantation life was abstract to Holder owing to his vast social and geographic distance from his enterprise, it was a fatal certainty to the humans and animals who experienced it.

That developments in markets, accounting, and finance made possible men like Holder helps to explain why, when poets and essayists on both sides of the Atlantic attacked slavery or animal cruelty, they often did so as part of their larger attacks on an emerging capitalism. Writers from a broad range of cultural and religious backgrounds, such as John Majoribanks (1759–1796), Lord Thomas Erskine (1750–1823), Percy Bysshe Shelley (1792–1822), Elizabeth Heyrick (1789–1831), Charlotte Smith (1775–ca. 1850), and Harriet Beecher Stowe (1811–1896), lambasted commerce in human and animal bodies, fitting biblical arguments about God's care for all creatures—human and nonhuman alike—into a "secular" discourse of enlightenment sensibility. They were far from identical in their critiques. But they were chosen for this study because of their common concern to make visible the invisible suffering of the vulnerable, as well as their warnings that to profit at another's expense was to become a beast-like "savage" oneself—a specter that might apply to individuals or entire societies. By confronting readers with human and animal suffering as well as the violence that caused it, these critics reflected and reinforced a key feature of enlightenment thought: the idea that sensory experience could shape, or even transform, the human self, including attitudes and consumer behaviors.

A historical outline of what we might call interspecies anticapitalism sheds light in a number of ways. It illuminates nineteenth-century mercantile dynamics that, in the minds of reformers, bound animals and enslaved people together across a huge geographic span. It shows how, by critiquing the race- and market-based forces that benefited some but exploited others, reformers created a discourse, however unintentional, around the ways human power structures intersected with the lives of animals. Among other things, this discourse can help explain the development of anticapitalist rhetoric in the animal rights movement. Further, it highlights possibilities for bringing together scholarly fields that have tended to remain separate. As historians of slavery have shown in recent years, the slave trade gave rise to new modes of finance, accounting, management, and political economy that in turn contributed to the development of industrial and free-market capitalism.[7] So, too, as environmental historians have shown, did Britain's push toward improved livestock husbandry generate capital that fed nineteenth-century markets.[8] If the development of capitalism forms a common link among these fields, then seeking to understand animals' and enslaved peoples' parallel experiences as capital can help us find in this scholarship the animal experience in human history, as well as the human experience in animal history. Such connections were apparent to early reformers like William Wilberforce, Fowell Buxton, Arthur Broome, and others, whose campaigns on behalf of animals and enslaved people often meant attacking the view that living things were mere commodities.

The danger that lurks in any such endeavor is to equate the experiences of animals and enslaved people, which were different in more ways than they were alike. Enslaved people learned to think of themselves as "black," endured sexual exploitation at the hands of their masters, fought back by sabotaging property (including, sometimes, their masters' livestock) and staging rebellions, and ultimately won their freedom. Livestock suffered physical brutality with no knowledge of "race" or how to stage collective revolt, often died before they could be used for breeding, and entered the stomachs of masters and enslaved people in the form of meat. The advocates whom this chapter traces seldom noted such distinctions. Yet their silence on this point, as well as their argument that mercantile logic itself flattened differences between animals and enslaved people, offers key insight into how race, animality, and capital helped constitute each other in England and its colonized territories in North America and the Caribbean. Nor did this dynamic end in 1850. Jennifer Marks's chapter on how epizootic disease ravaged Chicago's workhorses in 1872 brings to light how these animals, too, were viewed as living units of production whose labor was essential to the city's market-driven economy. Susan Nance's chapter highlights how ranchers in 1880s Montana treated livestock as simultaneously essential and expandable. Other authors in this volume echo that theme, and in some ways this chapter is a prologue to theirs. For even as early reformers sometimes blurred the experiences of human and animal victims, they also forwarded a vision of imperial England as a network of regions linked by dependence on and exploitation of living things.

THE PLANTATION MACHINE

As the House of Commons was debating William Wilberforce's slavery abolition bill in 1792, reformers waged an antislavery campaign that, since 1787, had flooded the presses with books, sermons, poems, and pamphlets.[9] Among those works was *Slavery: An Essay in Verse* (1792). Penned by John Majoribanks, a former soldier in the British army who had been stationed outside Kingston, Jamaica, from 1783 to 1787, the poem does not appear to have been widely read in its time.[10] But it is notable for its fierce salvos against the "curs'd commerce" of the slave trade, as well as its use of eyewitness accounts to bring readers inside the savage workings of the Jamaican sugar economy—both strategies that later antislavery and anticruelty reformers would adapt for their own purposes.[11]

Majoribanks's primary attack centered on the unnaturalness and inhumanity of the plantation machine. Addressing "Planters, Merchants, and others concerned in the Management or Sale of Negro Slaves," Majoribanks denied that "in the sight of God, any human being can be the *property* of another" (11). Enslavers had not only upset God's design but also sacrificed African bodies on the altar of mammon: "Av'rice only is their lord of all! / To him their rites incessantly they pay; / And waste for him the Negro's life away!" (8).

Furthermore, he argued, the efficiency-driven managerial schemes that treated humans as property both dehumanized enslaved people and transformed their masters into brutes. This much was clear from the actions of planters and overseers who hired enslaved men and women for task labor "but have no in'rest in them to preserve; / And if they labour, care not how they starve"; who "turn[ed] out old, or unserviceable slaves" to fend for themselves; and "who, with curs'd accuracy, count the days, / The hours of labour pregnancy delays."[12] So prevalent was the objectification of enslaved people as property, he recalled, that he once overheard a "bargain" that involved a "feotus" being "sold from the mother in whose womb it lay." Such "brutal ravishers" of African minds and bodies were capable of horrific cruelty. When one "gentleman" was absent, Majoribanks reported, his overseer cut off the leg of a tailor who tried to run away on the grounds "that as *he was a taylor*, the *property* was not a bit less valuable."[13]

This latter anecdote illustrated what Majoribanks understood as a key moral problem: enslaved peoples' suffering was invisible, even to those who inflicted it. Planters, for their part, often did not see their enslaved men, women, and children, particularly on plantations in Jamaica, where absenteeism rates were much higher than those elsewhere in the Caribbean.[14] Even as they lived in "luxury" on the other side of the Atlantic, they were blind to the pain and death on which that comfort was built: "What pangs," Majoribanks lamented, "what bloodshed, buy those joys for you! / Your injur'd slaves, perhaps, you *never saw*." While among the planter class were "undoubtedly many men of integrity," he thought, their comfortable lives in England were made possible by "speculators in human blood."[15] Even that speculation, though, to the extent it reduced humans to numbers, distanced those who practiced it from the possibility that enslaved people could feel pain or suffer death. A particular example was enslavers' practice of calculating whether to buy less human "property" up front and drain it quickly of life to make a quick profit, or to purchase a greater number of chattels as a longer-term investment:

> "Whether shall we," those precious scoundrels say,
> "Grasp Fortune quickly, or make long delay?
> "A hundred slaves we have no fund to buy;
> "The strength of *half that number* let us try,
> "With *mod'rate toil*, from practice it appears

"These slaves might live, perhaps, a dozen years;
"To us, you know, the matter will be even,
"If we can make as much of them in seven."

Majoribanks noted that he "repeatedly heard calculations made on this subject, with all the coolness and accuracy of an innkeeper estimating the probable expenditure of his post-horses." And while he granted that mules and horses were often better fed than enslaved people in Jamaica, he still "often heard planters, talking of their negroes, very gravely style them their *Cattle*."[16]

In making his case against the plantation enterprise, Majoribanks aimed to ameliorate the plight of enslaved people, not the plight of animals. He accepted that London's horses were treated badly and used this commonplace to argue that humans should not likewise suffer. Nor should we expect otherwise. By eliding animal suffering, Majoribanks fell within the mainstream of late eighteenth-century antislavery reform. But the larger themes he traced—planters' use of mercantile logic to define enslaved people as expendable property, as well as the ways geographic distance often separated enslavers from the pain that designation caused—would soon be used by others as a way to address animal and interspecies exploitation.

INTERSPECIES EXPLOITATION

Majoribanks's references to horses and other livestock were not lost on Holder's father, Rev. Henry Evans Holder II, who responded to Majoribanks that same year in *Fragments of a Poem*. That "baleful wit," he averred, "should remember that we are interested not merely in the *lives*, but in the *offspring* of our Negroes, which is not the case with our *mules* and *horses*," and that "the greatest brutes among the superintendents of the Negroes in the West-Indies, are Europeans," not Englishmen.[17] Yet even as Holder tried to distinguish enslaved people from animals and uphold England's claim to civility, observers in London had begun to notice that the same mercantile logic that shaped the lives and deaths of enslaved people also shaped the lives and deaths of the city's horses.

As industry boomed in London during the first half of the nineteenth century, so, too, did the demand for horses who could haul people and goods through the congested streets. And as firms competed for profit, time increasingly meant money. Horses found themselves pushed to move faster and work longer hours, often on as little food as necessary, which led to pain, starvation, and early death. At times, aged or decrepit horses who had worked all day were also forced to labor at night as garbage carriers, hidden from public view and thus likely to experience some of the worst treatment. But the length of horses' lives, like that of enslaved people, was deemed less important than the ability of a master to save money or extract profit. Thus, in keeping with a common criticism of plantations, an operating principle on London's streets was to use up horses quickly—many collapsed and died from sheer exhaustion—rather than invest money in treating them well or allow them rest in order to ensure a longer life.[18]

Like antislavery advocates, anticruelty reformers found themselves pushing against the defining structures of an emerging capitalism, including private property, speculation, market-driven production, and the use of finance and accounting to maximize profits.[19] Lord Thomas Erskine's efforts on behalf of London's animals is illustrative. In England, common law dictated that livestock were chattels, which meant that when Erskine argued in support of antibullbaiting bills in 1801 and 1802, he opened the door for members of Parliament like William Windham to remind landed interests that Parliament was trying to infringe on their private property rights. The opposition defeated both bills, and it was not until Parliament passed the Cruelty to Animals Act of 1835 that bullbaiting was outlawed in England.[20] In the meantime, Erskine employed religious language. Admitting in his 1809 anticruelty bill that "there can be no law for man" against animal abuse, he adapted the arguments of earlier reformers to warn that human "dominion" over nature was a God-given "moral trust" that, once broken, would dissolve humans into beasts. In addition, he took issue with the practice of treating old horses as assets to be depreciated, whereby the animals' usefulness was calculated based on how long they were expected to live: "I allude to the practice of buying up horses when past their strength, from old age or disease, upon the computation (I mean to speak literally) of how many days torture and oppression they are capable of living under, so as to return a profit with the addition of the flesh and skin, when brought to one of the numerous houses appropriated for the slaughter of horses." He charged that this practice was "committed under the deliberate calculation of intolerable avarice."[21] Thus, as Erskine pointed out, the issue was not a desire of profit per se, but a desire for profit at all costs, which led humans to treat horses as things that could be quantified, used up, and discarded. In a related way, Erskine had no problem placing horses among the nonhumans that were "constructed for [human] use."[22] But to reduce them to the status of beasts, devoid of physical or emotional sense, was to twist human dominion into outright abuse.

Another, more radical, option that was not available to MPs even had they wanted to pursue it was to tear down a system that elevated profit above life. The Romantic poet Percy Bysshe Shelley took this approach in *Queen Mab* (1813), a revolutionary poem in which a fairy, Queen Mab, leads the soul of Ianthe on a tour of the world's past, present, and future. Written for close friends, the poem caught the eye of an opportunistic bookseller who, a year before Shelley's death, distributed pirated editions through the black market. Between 1821 and the 1830s, publishers printed over a dozen such editions for working-class audiences in Britain and North America.[23] In so doing, they illustrated Shelley's central argument that everything was now a commodity. "All things are sold: the very light of heaven / Is venal," he lamented, even "the smallest and most despicable things / That lurk in the abysses of the deep."[24] According to Shelley, slavery and violence against nature and animals were the inevitable outcome of this perpetual "commerce," this "poison-breathing shade" that opened "the doors of premature and violent death" and "set the mark of selfishness, / The signet of its all-enslaving power / Upon a shining ore, and called it gold." The "tyrants" who benefited from this system acquired wealth "by the sale of human life," and all who worked within it, regardless of their degree of freedom, were "puppets," "slaves," "living pullies of a dead machine."[25] Animals shared in this suffering, as Shelley discussed in an essay on vegetarianism that was attached as notes to *Queen Mab*. But he also envisioned a utopia where humanity "no longer now"

> slays the lamb that looks him in the face,
> And horrible devours his mangled flesh,
> Which still avenging nature's broken law,
> Kindled all putrid humours in his frame,
> All evil passions, and all vain belief,
> Hatred, despair, and loathing in his mind,
> The germs of misery, death, disease, and crime.[26]

While Shelley was less forgiving of modern society than Lord Erskine, both men agreed that fitting society to the demands of markets was corrosive to civilization, and both adopted the strategy of confronting their audiences, even shocking them, with graphic proof of the decay. Erskine described the "most atrocious system of torture," which awaited London's horses when they grew too weak to labor and were taken to slaughterhouses. Rather than receiving the mercy of outright death, the horses were "literally starved to death, that the market might be gradually fed. The poor animals, in the mean time, being reduced to eat their own dung, and frequently gnawing one another's manes in the agonies of hunger."[27] Shelley, too, described human workers as starving "drones" and their masters as "gilded flies" who "feed / On the mechanic's labour." He perceived the greed-induced tragedy of

> yon squalid form,
> Leaner than fleshless misery, that wastes
> A sunless life in the unwholesome mine,
> Drags out in labour a protracted death,
> To glut [the masters'] grandeur; many faint with toil,
> That few may know the cares and woe of sloth.[28]

Decades before the founding of the Vegetarian Society in 1847, Shelley, along with others like John Frank Newton and William Lambe, were early exponents of a worldview in which meat-eating and violence went hand in hand. Shelley differed from his contemporaries, however, in his view that diet had political as well as personal repercussions. An individual's willingness to kill and eat a sentient animal found its analog in a state's readiness to torture, enslave, and treat as disposable the human laborers on which its wealth depended.[29] By uplifting the personhood of the soon-to-be-slaughtered lamb and describing human laborers in wretched, bestial terms, Shelley challenged the idea that humans were naturally superior to animals. At the same time, he called to account whom he took to be the true beasts: the merchants and profiteers who made violence possible.

Shelley formed part of an early Romantic anticapitalist tradition that included other poets and writers such as William Blake, John Keats, and Thomas Carlyle, among others, and that viewed mercantile interests as dangerous precisely because of the wide net of exploitation they cast over humans and nature.[30] In his poem *Isabella, or the Pot of Basil* (1818), Keats, perhaps anxious about his own lower-class status, took aim at "ledger-men" for whom romantic love ran counter to the logic of profit—a logic that ensnared both humans and animals. Isabella, a young woman whose brothers intend her to marry "some high noble and his olive trees," instead falls in love with Lorenzo, who works for her brothers. To compel Isabella to marry a wealthy man, the brothers murder Lorenzo. But Keats, echoing Majoribanks's criticism of absentee planters, is careful to show that they were already murderers from a distance. For they had been "enriched from ancestral merchandize," ill-gotten from "weary hand[s]" that labored in "torched mines," "noisy factories," and "in blood from stinging whip."[31] Though physically distant from their productive enterprises, they are nevertheless morally responsible. In stanza XV, which recalls Shelley's passage on the mines, Keats brings readers into these hidden places where humans and animals were suffering and dying "for them" within a vast production machine:

For them the Ceylon diver held his breath,
　And went all naked to the hungry shark;
For them his ears gush'd blood; for them in death
　The seal on the cold ice with piteous bark
Lay full of darts; for them alone did seethe
　A thousand men in troubles wide and dark:
Half-ignorant, they turn'd an easy wheel,
　That set sharp racks at work, to pinch and peel.[32]

To a certain extent, the parallels that Shelley and Keats drew between exploiting humans and hunting seals mirrored the parallels that Romantic poets drew between enslaving humans and caging birds during the same period.[33] But Shelley and Keats were concerned less with analogy and more with the underlying economic structures that linked human and animal exploitation and revealed them as originating in a common process. While they were not politicians in an official sense, they hoped to create social and political change through their writings.[34]

Nor were they alone. Even before *Queen Mab* and *Isabella* were written, antislavery poets critiqued the mercantile structures that made possible the exploitation of enslaved people, often incorporating arguments about the proper treatment of animals as well. In 1804, Charlotte Smith, an accomplished poet and writer whose work drew praise from Wordsworth, Cowper, Walpole, and others, explored the tragic consequences of treating animal and human lives as property in "To the Fire-Fly in Jamaica, Seen in a Collection." The poem juxtaposed the body of a firefly pinned for scientific study with the plight of an enslaved man:

> The recent captive, who in vain,
> Attempts to break his heavy chain,
> And find his liberty in flight;
> Shall no more in terror hide,
> From thy strange and doubtful light,
> In the mountain's cavern'd side,
> Or gully deep, where gibbering monkies cling,
> And broods the giant bat, on dark funereal wing.
>
> Nor thee his darkling steps to aid,
> Thro' the forest's pathless shade,
> Shall the sighing Slave invoke;
> Who, his daily task perform'd,
> Would forget his heavy yoke;
> And by fond affections warm'd,
> Glide to some dear sequester'd spot, to prove,
> Friendship's consoling voice, or sympathising love.[35]

Although the scientist's acquisitive grasp had extinguished the insect's life, Smith reestablished its beauty, as well as its relationship with the enslaved man. But both, in the end, were trapped, as Smith, too, found herself trapped in an unhappy marriage to a West Indian merchant's son and thus benefited from the very system of capital production she despised.[36]

Indeed, Smith exemplified a central anxiety that reformers exploited: the fear that humans would become less human, less civilized, if they too closely associated with a system that treated other lives as property and whose only purpose was to reap wealth for their masters. James Montgomery, for example, in his poem "The Ocean" (1805), lamented "the poor disinherited outcasts of man, / Whom Avarice coins into slaves," and portrayed the "port" at which they arrived after their transatlantic voyage as a place

> Where the vultures and vampires of Mammon resort;
> Where Europe exultingly drains
> The lifeblood from Africa's veins;
> Where man rules o'er man with a merciless rod,
> And spurns at his footstool the image of God![37]

The corollary to rejecting one's status as a being made in "the image of God" was the loss of humanity itself, so that one might become an animal enslaved to cruel appetites. To guard against this possibility, Isabella Oliver warned in "On Slavery" (1805), children should be taught neither to "give the meanest insect useless pain" nor "trample down, the poor defenceless black," lest

> God's image in his creature they deride,
> And daily grow in indolence and pride,
> With ignorance and cruelty combin'd;
> A Slavery of the most ignoble kind!

Still, Oliver was not hopeful: "How many futile reasons have been given, / For mixing God and mammon, sin and heaven!"[38]

As Smith, Montgomery, and Oliver made clear, the contours of early anticruelty thinking revolved as much around upholding humanity and preventing animal-like behavior as it did around extending kindness to animals. Advocates emphasized different aspects of this discourse depending on species and context. Children should be kind to all creatures, whereas adults should be merciful to the horses, cattle, and other livestock with whom they most often worked. In this way, early nineteenth-century advocates were already outlining the terms of a later pro-animal tradition, which revolved—and continues to revolve—around relationships among "humans," "nonhumans," and "beasts," and which was selective about which animals were worth protecting.

What made matters difficult for reformers was the fact that people did not have to give reasons for "mixing God and mammon."[39] They had only to consume the goods that enslaved people produced in order to be complicit in their own dehumanization and that of others. As Joshua Marsden put it in "The Sale of Slaves, or a Good Bargain" (1810),

We have slav'd the human race,
Sunk the mortal to a brute;
Tumbled manhood from its place,
To get sugar, rum, and fruit.

Sweet our coffee, sweet our tea,
But in bitterness of soul,
Many a wretch has pin'd away,
To ameliorate the bowl.[40]

In this way, people's own buying habits, which in a market-driven system fueled the demand for enslaved laborers, would bring to pass what Majoribanks, Shelley, Keats, and others feared: a greedy and violent world that dehumanized enslaved people and enslavers alike.

SEEING PAIN

As London increased in size over the course of the eighteenth and nineteenth centuries, its people required more and more animals to be slaughtered for meat. Similarly, expanding populations in America and throughout Europe required ever larger numbers of enslaved people to produce the tobacco, sugar, cotton, and other goods on which these populations depended. Humanitarian writers during this period had a keen sense for the relationship between growing consumer demand and the suffering it caused. By making visible the suffering of animals and enslaved people, these writers sought to collapse the distance between producers of goods and services and those who blindly consumed them. In so doing, they exemplified the common idea during this period that seeing, as well as how you interpreted what you saw, helped constitute who you were and what you would become.[41] Sight could be an instrument of corruption or compassion, savagery or civilization. Part of the work of reform, then, was to transform society by confronting it with images of what it had become. To accomplish this, Majoribanks took readers inside the cold logic of the Jamaican plantocracy. Keats in *Isabella* unveiled the machinations of Isabella's brothers. Shelley's poetic eye, enlightened by an all-seeing fairy, perceived "the unwholesome mine," the "abysses of the deep," and the man who killed "the lamb that looks him in the face," regardless of what hid these things from ordinary human sight. Erskine spoke of violence against horses in plain sight, but he also brought his audience into the slaughterhouse—itself an eighteenth-century development that was meant to hide suffering, to remove it from the public eye.[42]

Aiming to change readers' attitudes toward animals, Elizabeth Heyrick, a British philanthropist and activist, attempted to see animals' pain in her 1823 essay on the "unrestrained cruelty" of Smithfield Market, where livestock were sold for slaughter in London. Inviting readers to view the market from "an upper window" that offered "a general view," Heyrick proceeded to describe how, at nine o'clock on Sunday evening, some 35,000 sheep and lambs "are driven into pens in the centre of the area," followed by "horned cattle"; and how the animals were then examined by "purchasers" and goaded by "wretches" until they bled. "I have heard them bellow," she recounted, "and have seen their eye-balls roll, as in intense anguish from violent blows." Yet "they generally exhibit the most patient endurance of every kind of persecution, giving few outward indications of suffering," though they endured blows to the eyes and face, circumcision, and the severing of their tails.[43]

In Heyrick's mind, animal cruelty in the imperial center, England, was linked to slavery along its peripheries. On the one hand, she doubted enslaved people were treated as badly as England's livestock: "We have heard much," she wrote, "of the barbarities and horrors of slavery . . . but, on no part of the globe . . . can I imagine scenes of more atrocious cruelty, of more fiend-like depravity, than those exhibited in Smithfield-market." On the other hand, she believed that cruelty knew no species bounds: "Our treatment of animals," she wrote, "may be regarded as an accurate criterion of our humanity towards our own species. Whoever, from motives of interest, vanity, passion, or caprice, subjects them to oppression or unnecessary suffering, will, from the same motives, oppress and afflict his own species, whenever he can do it with impunity." So it was that "an American republican . . . disclaims his relationship to his sable brethren—considers them as creature of an inferior species . . . whom he may, with impunity, buy and sell, and lash as beasts of burden."[44] In this way, Heyrick picked up the refrain of poets like Oliver, Montgomery, and Smith, who viewed the boundary between humans and animals as porous. But while these earlier critics highlighted how cruelty might make beasts out of masters, Heyrick noted its double edge: it might also make beasts out of enslaved people.

Masters, too, ought to beware of cruelty's effects. Heyrick's attacks on the economics of animal cruelty centered not on reducing animals to a set of numbers, treating them as assets, or depreciating them over time. Rather, they revolved around the idea of "selfishness." A person who gave in to selfishness would inevitably acquiesce to "cruelty," which would in turn lead him to become less human. As an example of how this worked, she used the familiar example of London's horses. Citing the "fashionable and almost universal system of rapid travelling" by horse across a growing city, she took aim at procedures that were painful to horses but that saved the owner or driver time and money. "He who can press the exertions of his horse beyond its strength," she charged, "can remorselessly sentence it to the torturing operation of firing, in order artificially to brace its relaxed and enfeebled muscles," or "to enhance its saleable value, subject it to the torture of nicking or docking, must be a selfish, sordid, unfeeling character."[45] "Firing" entailed burning or cauterizing parts of a horse's leg in order to more quickly heal injuries, "nicking" meant cutting certain tail muscles in order to make the tail stand up," and "docking" meant cutting off part of the tail. Such practices reflected not just an individual but a "national cruelty" that was "an inevitable result of national selfishness."[46] This selfishness stemmed primarily from a defect of sight. To see individual profit but not the equine suffering on which it depended was to sever one's connection with living things, and ultimately to lose oneself in the very act of trying to serve oneself.

In seeing that England was selfish toward animals, Heyrick also saw, as had Jeremy Bentham six years earlier, a parallel in the political and economic realm: aristocratic abuse of the working classes. Heyrick decried how England's "national sin" toward horses manifested itself as well in the poor treatment "the labouring classes," who "are laid under the highest possible exactions in rent, taxes, &c." through which "we strive to obtain the various productions of their industry for the lowest possible remuneration."[47] In his *Plan of Parliamentary Reform* (1817), Bentham, too, in order to show the possibility that the English monarch did not have poor people's interests at heart, drew parallels to horses and enslaved people. "The Slave-holder," he reasoned, "has an interest in common with his slaves . . . and so has the "Mail-Coach Contractor in common with that of his horses." Yet that interest arose not from the inherent value of enslaved people or horses, but from their use value. For, "while working them, and so long as they appear able to work, he accordingly allows them food. Yet . . . notwithstanding this community of interest, so it is that . . . too often

Negro as well as horse are worked to the very death. —How happens this . . . but because in the same breast with the conjunct interest is lodged a separate and sinister interest, which is too strong for it." "The condition of the poor people," Bentham warned, "is day by day approaching nearer and nearer to the condition of the Negro and the horse."[48] Thus, Bentham was aware of how humans and animals experienced mutual, but at times unequal, vulnerabilities in a capitalist, imperial England. Still, the same "sinister interest," or profit-driven selfishness, that guided post-horse drivers on city streets was also operating in urban slums and distant plantations: human and animal exploitation were twins. Heyrick understood these developments as unnatural. In her Smithfield pamphlet, she cautioned that God would judge cruel masters and that he saw all the inner workings of the market: the "sufferings of the animals . . . cannot escape His observation." In keeping with her emphasis on the transformative power of sight, she recommended two practical reforms: an end to "cattle . . . be[ing] driven into the market before day-light, it being suspected that much barbarity is practised under cover of night," and an increase in the number of "inspectors" who could keep an eye out for cruel practices.[49]

The movement toward direct action in response to visible interspecies cruelty grew during the first part of the nineteenth century. The formation of the Society for the Prevention of Cruelty to Animals (SPCA) in 1824 was an important step toward reform, and the idea that human and animal interests were bound together was apparent to its supporters. William Wilberforce and Fowell Buxton, both founding members of the SPCA, also supported campaigns to end slavery. In its 1824 *Prospectus* and its 1829 *Objects of Address*, the society justified its existence as an extension of England's "tender care for our suffering brethren, of every colour and complexion." And extending Heyrick's and Bentham's concern for working and enslaved people, it asserted that animals constituted a "class" that suffered "in a ten-fold degree" compared to the "poor," "blind," and "dumb," since, in the view of the SPCA, they were all of these things at once, and thus suffered "the most abject slavery" possible. By 1829, the society could point to a number of "objects effected or attempted," including the amelioration of animal cruelty in London's streets and Smithfield Market.[50] In 1838, fifteen years after advocating greater oversight at Smithfield, Heyrick adopted a direct market-based mode of reform in a pamphlet that called for "immediate" abolition of slavery.[51] Whereas she did not urge readers to boycott Smithfield meats in 1823, she did exhort readers to boycott West

Indian goods produced by enslaved people in 1838. "The planter refuses to set his wretched captive at liberty," she lamented, "treats him as a beast of burden, compels his reluctant unremunerated labour under the lash of a cart whip, why? Because *we* furnish the stimulant to all this injustice, rapacity, and cruelty, by *purchasing its produce*." The remedy was "*abstinence from the use of West Indian productions, sugar* especially. . . . When there is no longer a market for the productions of *slave labor*, then, and not *till then*, will the slaves be emancipated."[52] Heyrick, like Majoribanks, Shelley, and Keats, understood the social and geographic distance that separated the sites of cruelty from the consumers who benefited from it, and now applied those observations in the pursuit of direct economic action. In this way, she used her own sight and called on others to use theirs as a means to combat market-based cruelty in England and abroad.

By the time Heyrick published her 1838 pamphlet, it was possible to see a discourse forming in which masters, enslaved people, consumers, and animals were entwined in a web of oppression that consisted of markets for sugar, meat, horse-drawn cabs, and other goods and services. These markets spawned competition, which in turn necessitated cost-saving measures, which then led to exploitation and suffering. All of this meant that markets produced violence, even if consumers were not always able to see their role in upholding that violence. Consumers might boycott plantation-grown sugar, as many chose to do between the 1790s and 1830s,[53] but they might also change their exploitative behavior by looking carefully at their own relationship to animals. It was this sort of looking that John Wesley, who died in 1791 but whose sermons continued to be reprinted throughout the nineteenth century, sought to encourage in his sermon on "general deliverance."[54]

Wesley's appeal centered on the idea that while humans and animals now suffered violence together, they would also be redeemed together. It followed that because God valued animals enough to bring them into heaven, humans had a moral duty to treat them well on earth. At the beginning of creation, Wesley argued, humans and animals were equal in that they both possessed "self motion, understanding, will, and liberty." The difference was that while humans were designed perfectly to understand God, the "lower creatures" were designed perfectly to understand humans.[55] But after the Fall destroyed perfect understanding, "man [was] deprived of *his* perfection, his loving obedience to God," and "brutes [were] deprived of *their* perfection, their loving obedience to man." So, too, were animals deprived of peace. Wild animals now killed each other, and domestic animals, who still retained some of their natural inclination to serve humans, endured "brutal violence," "outrage," and "abuse" on the part of humans. Pointing especially to horses and dogs, Wesley exclaimed:

> What returns for their long and faithful service do many of these poor creatures find! And what a dreadful difference is there, between what they suffer from their fellow brutes, and what they suffer from the tyrant man! The lion, the tiger, or the shark, give them pain from mere necessity, in order to prolong their own life; and put them out of their pain at once: but the human shark, without any such necessity, torments them of his free choice; and perhaps continues their lingering pain, till, after months or years, death signs their release.[56]

Yet Wesley argued that these suffering animals would share in the salvation of the world. Citing the passage in the book of Romans that reads, "They themselves also shall be delivered . . . from the bondage of corruption, into [a measure of] the glorious liberty of the children of God," Wesley imagined a heaven populated by perfected humans and animals. To those who objected that animals served no real "use," Wesley rejected the notion that a living being could be reduced to its use value and affirmed that God could see in these creatures what humans could not: "Consider how little we know of even the present designs of God; and then you will not wonder, that we know still less of what he designs to do in the new heavens and the new earth."[57]

Wesley's attempts to see social problems as he supposed God saw them, and to use that vision to promote reform, were extended by Christian abolitionists in North America such as Charles Finney, William Lloyd Garrison, Theodore Weld, and Sarah and Angelina Grimke. In 1837, Weld published *The Bible against Slavery*, which, as its title suggests, used biblical passages to discredit slavery. Two years later, the Grimke sisters, who had left their father's plantation in South Carolina, converted to Quakerism, and taken up the abolitionist cause in Philadelphia, joined Weld in writing *American Slavery as It Is: Testimony of a Thousand Witnesses* (1839), which sold more than a hundred thousand copies in the year it was published.[58] Through a combination of firsthand testimony and articles from southern newspapers, *American Slavery* offered a lacerating critique of enslavers and highlighted how, in a system where humans and animals were capital assets, all that mattered was their use value.

In an attempt to convince readers of the horrific abuse that unfolded on plantations, the Grimkes and Weld argued that because masters defined enslaved people as "property, and not as *persons*," they could mete out punishments that "if inflicted upon whites, would fill [them] with horror and indignation." Enslavers viewed enslaved men and women as "domestic animals," "chattels and working animals," "dogs," and "merchandise."[59] This latter category made clear that these were no empty metaphors: the logic of capital acquisition bound animals and enslaved people together as chattels in a market-based system, which in turn dictated how masters would treat them. Enslavers gave slaves "names which . . . are the same and similar to those given to their horses and oxen," tracked their "increase" as they would that of "flocks and herds," and called enslaved women "breeders."[60] The authors cited an abolitionist speech to the Virginia Legislature in 1832, in which James Gholson attributed these designations to enslaved people's status as chattels. Just as "the owner of *brood mares*," he explained, had a "[reasonable right] to their product," so "the owner of *female slaves* [had a reasonable right] *to their increase*." And while it might be expensive for an owner to raise an enslaved person from infanthood, "the value of the property justifies the expense." The "property" was valuable not only because slave prices were rising in response to demand in the cotton states during the 1830s, but because state laws enabled masters to mortgage enslaved people, sell them, and treat them as transferable goods just as they would animals.[61] In America as in England, greed fueled the system: "Avarice alone can drive . . . this *infernal* traffic," and enslaved women and men were the "wretched victims of it, like so many post-horses, *whipped to death* in a mail coach."[62]

Abolitionists such as the Grimkes and Weld were concerned with emancipating enslaved people, not with interspecies justice. Indeed, during the early part of the nineteenth century, American antislavery writers tended to portray animals as foils for the poor treatment of enslaved men and women rather than as beings who deserved protection in their own right. But in terms of the relationship of animals and enslaved humans to capital, there could be no mistake in the abolitionist message: once defined as property, human and animal chattels suffered and died together within a system that was only capable of viewing them as more or less profitable resources. In the process, enslaved and enslavers alike lost their humanity, so that the nation was filled with animals.

In her novel *Uncle Tom's Cabin* (1852), which sold three hundred thousand copies in North America and one million copies in England the year it was published, Harriet Beecher Stowe exemplified these concerns. She particularly explored them in the relationship between Tom, an enslaved man, and Simon Legree, a cruel planter. Legree's campaign against the humanity of his enslaved was rooted in his definition of them as property, which, he believed, entitled him to dehumanize and kill them however and whenever he wished. Indeed, Legree was a man for whom the dehumanization of enslaved men and women was a matter of everyday business, a fact as easily borne out by his proud manufacture of Sambo and Quimbo into things "lower . . . than animals" in nature as it was by his fervent desire to "break" the will of Tom, force him to "beg like a dog," and finally reduce him to ashes by means of a "slow fire."[63] Even as Legree wanted Tom to behave like an animal, he wanted his animals to behave like killers. For as killers, they performed the useful function of intimidating enslaved people into working harder. As Legree explained, he was as receptive to the idea of feeding enslaved men and women to his "bull-dogs" for supper as he was to burning them to death, and he even managed to kill or let die with a certain amount of industrial pride: "I don't go for savin' n——s. Use up, and buy more, 's my way. . . . When one n——'s dead, I buy another; and I find it comes cheaper and easier, every way."[64] The market, then, was for Legree a tool for dehumanization and a natural—and for him financially beneficial—outworking of it. For even as the market enabled him to view enslaved men and women as dollar amounts rather than humans, so the presence of that market depended on a set of laws and practices that defined enslaved people as buyable, sellable, and replaceable property. Within the context of Stowe's narrative, then, market-based capitalism was intimately connected with Legree's tyranny over human bodies and souls. Indeed, the omnipresent death and decay of his farm were merely the outward symbols of "souls crushed and ruined, evil triumphant, and God silent."[65] At the same time, they furnished evidence that Legree was the quintessential miser, the cold and calculating individual who hoarded money to suit his own avarice, not to better himself or society.

As in Wesley's portrayal of the cruel master as a "human shark," however, Stowe upset Legree's pretense to superiority. For she reasserted Tom's free will and his eternal value, both uniquely related to his humanity, against Legree's attempts to beat these things out of him. In the main sequences relating to

Tom's beating and death, Stowe variously referred to Legree as a "ferocious beast," a "dog," an "alligator" or "rhinoceros," and an "incensed lion . . . foaming with rage." At one point, Stowe drove home the man-beast contrast by appropriating a passage from the book of Daniel to Tom's plight: "The Lord God hath sent his angel, and shut the lion's mouth."[66]

In the meantime, reformers on both sides of the Atlantic made gradual progress. On the English side, the Slavery Abolition Act that came into effect in 1834 freed more than eight hundred thousand enslaved people in the Caribbean, South Africa, and Canada, and made the trade in human bodies illegal throughout the British Atlantic World. North America's enslaved population, which totaled more than two million in 1830 and grew to nearly four million by 1860, would not gain freedom until after the Civil War. Reforms on behalf of animals kept roughly apace. In England, the 1835 Cruelty to Animals Act remained animals' only substantial legal protection until Francis Power Cobbe established the National Anti-Vivisection Society in 1875, and it was not until 1892 that Henry Salt published his landmark essay, *Animals' Rights Considered in Relation to Social Progress*. In America, anticruelty reformers won legislative successes in New York (1829), Massachusetts (1836), and Philadelphia (1855), but widespread campaigns on behalf of animals would not begin until after the founding of the American Society for the Prevention of Cruelty to Animals in 1866.[67] A broad question that emerges from this chronology is why, even after decades of antislavery efforts and anticruelty reforms, enslavement of humans and indifference toward animals continued to exert a hold—as they still do today. While the answer to this question is beyond the scope of this chapter, humanitarian advocates clearly understood that as long as society was profiting from the commodification of animals and enslaved people, and as long as consumers remained distant from the suffering that resulted, interspecies exploitation would continue in England and its peripheries.

Whether reformers invoked religious arguments, or passed laws and regulations, or marshaled broader humanitarian arguments in their attacks on slavery and cruelty to animals, they often pointed to the ways that market-based forces created moral distance between those who had power and those who did not. Speculating on, depreciating, and valuing living beings according to prices dictated by markets had the effect of abstracting vulnerable humans and animals, which in turn reduced masters'—and consumers'—sense of responsibility for the well-being of those lives. Nor were market forces capable of generating their own ethical critique. When antislavery and anticruelty reformers wrote against the effects of an emerging capitalism, they did so in ways that reduced moral distance through sight and bound together suffering humans and animals. The act of seeing pain and suffering, and of writing in ways that enabled their audiences to see these things, had the power to recast human and animal lives not as capital assets but as sentient beings. Taken collectively, humanitarian reformers' large, diffuse, sometimes contradictory body of work, which this essay has only sampled, is an invitation to probe historical links between enslaving humans and managing animals, and the implications of commodifying both.

NOTES

1. *Minutes of the Society for the Improvement of Plantership in the Island of Barbados* (Liverpool: Thomas Kaye, 1811), 66.

2. Aristotle held that just as it was natural and good for certain animals to labor for humans, so was it natural and good for certain humans to labor in slavery. These "slaves by nature," as he called them, had bodies that were suited to work, whereas their masters had bodies suited for "community life." Aristotle, *Politics*, quoted in Thomas Wiedemann, *Greek and Roman Slavery* (London: Croom Helm, 1981), 18–19, 17. For the idea that Europeans used Aristotle's notion of "natural slavery" as a justification for enslaving Africans, see Karl Jacoby, "Slaves by Nature? Domestic Animals and Human Slaves," *Slavery & Abolition* 15, no. 1 (1994): 95; Marjorie Spiegel, *The Dreaded Comparison: Human and Animal Slavery*, rev. ed. (New York: Mirror Books, 1996), 73–75; Mark S. Roberts, *The Mark of the Beast: Animality and Human Oppression* (West Lafayette, IN: Purdue University Press, 2008), ch. 5.

3. Caitlyn Rosenthal, *Accounting for Slavery: Masters and Management* (Cambridge, MA : Harvard University Press, 2019), 42.

4. Will of Henry Evans Holder, Jan. 30, 1816, 8973/2, Bristol Records Office.

5. In the nineteenth-century Caribbean and US South, using animals and enslaved people as "debt securities" often meant mortgaging them (i.e., putting money down at purchase and paying off the balance over time) or using them as collateral for loans. Owners could sell or transfer these assets during life or, as in Holder's case, at death. See Bonnie Martin, "Slavery's Invisible Engine: Mortgaging Human Property," *Journal of Southern History* 76, no. 4 (2010): 817–66; Christer Petley, "Managing 'Property': The Colonial Order of Things within Jamaican Probate Inventories," *Journal of Global Slavery* 6, no. 1 (2021): 81–107.

6. Elizabeth Abbot, *Sugar: A Bittersweet History* (New York: Penguin, 2008); Richard S. Dunn, *Sugar and Slaves: The Rise of the Planter Class in the English West Indies, 1624–1713* (Chapel Hill: University of North Carolina Press, 1972), 301.

7. See, e.g. , Rosenthal, *Accounting for Slavery*; Sven Beckert and Seth Rockman, eds., *Slavery's Capitalism: A New History of American Economic Development* (Philadelphia: University of Pennsylvania Press, 2016); Edward Baptist, *The Half Has Never Been Told: Slavery and the Making of American Capitalism* (New York: Basic Books, 2014); Sven Beckert, *Empire of Cotton: A Global History* (New York: Vintage Books, 2014); Walter Johnson, *River of Dark Dreams: Slavery and Empire in the Cotton Kingdom* (Cambridge, MA: Harvard University Press, 2013).

8. Rebecca J. H. Woods, *The Herds Shot Round the World* (Chapel Hill: University of North Carolina Press, 2017); Richard Perren, *The Meat Trade in Britain, 1840–1914* (London: Routledge, 1978); Robert Trow-Smith, *A History of British Livestock Husbandry*, vol. 2, *1700–1900* (London: Routledge, 1959).

9. John R. Oldfield, *Popular Politics and British Anti-Slavery: The Mobilisation of Public Opinion against the Slave Trade, 1787–1807* (London: Routledge, 2017).

10. Karina Williamson, "The Antislavery Poems of John Majoribanks," *EnterText* 7, no. 1 (2008): 61, 63.

11. John Majoribanks, *Slavery: An Essay in Verse* (Edinburgh, 1792), 10. On the relationship between abolitionism and capitalism more generally, see Seymour Drescher, *Capitalism and Antislavery: British Mobilization in Comparative Perspective* (New York: Oxford University Press, 1987).

12. Majoribanks, *Slavery*, 11, 8, 21, 12, 13.

13. Majoribanks, *Slavery*, 21, 22, 15.

14. Rosenthal, *Accounting for Slavery*, 42.

15. Majoribanks, *Slavery*, 18, 26.

16. Majoribanks, *Slavery*, 17–18, 22.

17. Henry Evans Holder II, *Fragments of a Poem: Intended to Have Been Written in Consequence of Reading Major Majoribanks's Slavery* (London, 1792), [13], 9–10, 11.

18. Hilda Kean, *Animal Rights: Political and Social Change in Britain since 1800* (London: Reaktion, 1998), 50–52.

19. On the reformist intersection between antislavery and anticruelty, see Kean, *Animal Rights*, 33–35; Keith Thomas, *Man and the Natural World: A History of the Modern Sensibility* (New York: Pantheon, 1983), 184–85.

20. Harriet Ritvo, *The Animal Estate: The English and Other Creatures in the Victorian Age* (Cambridge, MA: Harvard University Press, 1987), 125–26, 128.

21. Lord Thomas Erskine, *Cruelty to Animals: The Speech of Lord Erskine in the House of Peers* (Edinburgh, 1809), 4, 2, 14.

22. Erskine, *Cruelty to Animals*, 3. On the complex distinction between "nonhumans" and "beasts" in a range of historical contexts, see Bénédicte Boisseron, *Afro-Dog: Blackness and the Animal Question* (New York: Columbia University Press, 2018), ch. 3.

23. Jen Morgan, "The Transmission and Reception of P. B. Shelley in Owenite and Chartist Newspapers and Periodicals," PhD diss., University of Salford, Salford, UK, 2014, 40–53.

24. Percy Bysshe Shelley, *Queen Mab* (New York: Frederick Campe, 1820), 33.

25. Shelley, *Queen Mab*, 30.

26. Shelley, *Queen Mab*, 52.

27. Erskine, *Cruelty to Animals*, 15.

28. Shelley, *Queen Mab*, 19.

29. Tristam Stuart, *The Bloodless Revolution: A Cultural History of Vegetarianism from 1600 to Modern Times* (New York: W. W. Norton, 2007), ch. 26.

30. Jerry W. Lee, "Capital, Class, and Representations of Isabel/la in John Keats's *Isabella*," *The Explicator* 70, no. 3 (2012): 183–86; E. P. Thompson, *Witness against the Beast: William Blake and the Moral Law* (Cambridge: Cambridge University Press, 1993); Michael Löwy, "The Romantic and Marxist Critique of Modern Civilization," *Theory and Society* 16, no. 6 (Nov. 1987): 891–904.

31. John Keats, *Lamia, Isabella, The Eve of St. Agnes, and Other Poems* (London, 1820), 56.

32. Keats, *Lamia, Isabella, The Eve of St. Agnes, and Other Poems*, 56.

33. For examples of these poems, see *British Women Poets of the Romantic Era*, ed. Paula R. Feldman (Baltimore: Johns Hopkins University Press, 1957), 89–90, 595–98, 736–38, 808–9, 838–40; Anna Laetitia Barbauld, "Epitaph on a Goldfinch," in *Works* (New York, 1826), 2:322–23.

34. Jeffrey N. Cox, *Poetry and Politics in the Cockney School: Keats, Shelley, Hunt, and Their Circle* (Cambridge: Cambridge University Press, 1998); Morris Dickstein, "Keats and Politics," *Studies in Romanticism* 25, no. 2 (Summer 1986): 175–81.

35. Charlotte Smith, "To the Fire-Fly in Jamaica, Seen in a Collection" (1804), in James Basker, ed., *Amazing Grace: An Anthology of Poems about Slavery, 1660–1810* (New Haven, CT: Yale University Press, 2002), 595.

36. Smith, "To the Fire-Fly."

37. James Montgomery, "The Ocean" (1805), in Basker, *Amazing Grace*, 611, 612.

38. Isabella Oliver, "On Slavery," in Basker, *Amazing Grace*, 618, 619.

39. Oliver, "On Slavery," 619.

40. Joshua Marsden, "The Sale of Slaves, or a Good Bargain" (1810), in Basker, *Amazing Grace*, 656.

41. Hilda Kean, *Animal Rights*, 27; Janet M. Davis, *The Gospel of Kindness: Animal Welfare and the Making of Modern America* (New York: Oxford University Press), 41.

42. Thomas, *Man and the Natural World*, 294–5; Amy J. Fitzgerald, "A Social History of the Slaughterhouse: From Inception to Contemporary Implications," *Research in Human Ecology* 17, no. 1 (2010): 58–69. On the ways that modern slaughterhouses manipulate what consumers and workers can see to mask violence, see Timothy Pachirat, *Every Twelve Seconds: Industrialized Slaughter and the Politics of Sight* (New Haven, CT: Yale University Press, 2011).

43. Elizabeth Heyrick, *Cursory Remarks on the Evil Tendency of Unrestrained Cruelty . . . in Smithfield Market* (London, 1823), 4, 5, 6, 7, 8.

44. Heyrick, *Cursory Remarks*, 3, 4, 14, 13.

45. Heyrick, *Cursory Remarks*, 16, 22, 15.

46. Heyrick, *Cursory Remarks*, 22.

47. Heyrick, *Cursory Remarks*, 22.

48. Jeremy Bentham, *Plan of Parliamentary Reform* (London, 1817), xxvi, xxvii.

49. Heyrick, *Cursory Remarks*, 11, 21.

50. Diana Donald, *Picturing Animals in Britain, 1750–1850* (New Haven, CT: Yale University Press), 224–25; *Objects and Address of the Society for the Prevention of Cruelty to Animals* (London, 1829), 6, 9–11.

51. Elizabeth Heyrick, *Immediate, not Gradual Abolition: Or, an Inquiry into the Shortest, Safest, and Most Effectual Means of Getting Rid of West Indian Slavery* (Boston, 1838).

52. Heyrick, *Immediate, not Gradual Abolition*, 4, 5.

53. Julie L. Holcomb, "Blood-Stained Sugar: Gender, Commerce and the British Slave Trade Debates," *Slavery & Abolition* 35, no. 4 (2014): 611–28; Clare Midgley, "Slave Sugar Boycotts, Female Activism and the Domestic Base of British Anti-Slavery Culture," *Slavery & Abolition* 17, no. 3 (1996): 137–62; Charlotte Sussuman, "Women and the Politics of Sugar, 1792," *Representations* (1994): 48–69.

54. John Wesley, "Sermon LXV: The General Deliverance," in *The Works of the Reverend John Wesley, A. M.* (New York: Waugh and Mason, 1833), 2:49–57.

55. Wesley, "Sermon LXV," 50, 51.

56. Wesley, "Sermon LXV," 52, 54.

57. Wesley, "Sermon LXV," 54, 56.

58. Sarah N. Roth, *Gender and Race in Antebellum Popular Culture* (New York: Cambridge University Press, 2014), 82.

59. *American Slavery as It Is: Testimony of a Thousand Witnesses* (New York, 1839), 110.

60. *American Slavery as It Is*, 110.

61. *American Slavery as It Is*, 182.

62. *American Slavery as It Is*, 173.

63. Harriet Beecher Stowe, *Uncle Tom's Cabin* (1852), ed. Ann Douglas (New York: Penguin, 1986), 493, 527, 538, 539.

64. Stowe, *Uncle Tom's Cabin*, 492, 495.

65. Stowe, *Uncle Tom's Cabin*, 552.

66. Stowe, *Uncle Tom's Cabin*, 508, 541, 582–83, 541.

67. Seymour Drescher, *Abolition: A History of Slavery and Antislavery* (Cambridge: Cambridge University Press, 2009), chs. 9, 10, 11; Kean, *Animal Rights*, chs. 2, 4; Davis, *Gospel of Kindness*, 28, 180.

JENNIFER MARKS

two Chicago's 1872 Equine Influenza Epizootic and the Evolution of Urban Transit Technology

Horses, quite literally, powered urban America throughout the nineteenth century. Cities relied upon equine bodies to pull streetcars, haul construction equipment, transport goods, and carry riders from place to place. The energy provided by horses fueled a complex, entangled, and ultimately vulnerable urban circulatory system.[1] Humans provided housing and care for the horses who drove the city; horses provided energy for the city as well as companionship and employment for the humans who fed and kept them. The two species crafted relationships that could simultaneously be compassionate and controlling, altruistic and abusive. The intricate and often unequal human-animal relationships that powered the urban environment are difficult to see. However, a three-week period in Chicago during the fall of 1872 provides us with an exceptional moment of clarity.

In late October of that year, an outbreak of equine influenza tore through the public and private stables of urban North America, debilitating horses with severe coughs and other respiratory tract symptoms. The virulent strain traveled along railroad routes, infecting tens of thousands of horses from Ontario to New York to Boston. The epizootic's disruption ironically illuminates and helps us disentangle the networks and consequences of Chicago's animal-powered urban ecosystem. By shutting down the city's street-level circulatory system, the epizootic (the animal equivalent of an epidemic) made animal power on a citywide scale newsworthy, leaving a record of an urban environment vulnerable to disease and an evolution of transportation technology that was animal as well as human.

Chicago felt the epizootic acutely, as much of the city was still recovering from the Great Fire. A great swathe of the city—more than three square miles—and hundreds of its horses had burned the year before. Now, with their motors in respiratory distress, Chicago's cabs, omnibuses, streetcars, and construction equipment halted. Men, women, and children walked. Companies purchased oxen to deliver goods and luggage between railroad depots. Even dogs were put to work. The city council called a special meeting to debate the use of the much-maligned "dummy" steam engines (locomotives disguised to look like passenger cars). Chicagoans were struck by the eerily silent city, with the sharp din of horseshoes, driver commands, and whinnying suddenly absent from the streets.

The growing metropolis stopped functioning and feeling like a city. But the city was not to be outdone in its suffering. "Chicago never does things by halves. We mean to be champions in everything, even in the epizootic line; . . . this city will be as much superior to other cities in that disease as it is to-day ahead of them on conflagrations," declared the *Tribune*.[2] Print media provided a space in which city officials, business owners, and the general public could express their appreciation for and anxieties about living in an animal-powered metropolis. At the same time, newspapers hosted debates about the best way to reckon with the city's new fragility.

Sean Kheraj describes turn-of-the-century animal cities as networked ecosystems that "share common characteristics and the capacity to behave as unified disease pools for both animals and humans."[3] Chicago was like other American cities during this period: the very urban fabric that made the city function came from horses vulnerable to disease. Rather than telling the unique experience of Chicago, this case study explores how a networked ecosystem functioned in the midst of an outbreak in order to explain the coevolution of urban technological development and systemic animal exploitation on a city-level scale. Like the stories told by Mary Trachsel and Susan Nance elsewhere in this collection, this is a tale about the unintended consequences of an urban ecology entirely dependent on animals. In this case, Chicago's dependence on animals bred technological development. The modern city did not evolve after animals—it evolved with and through them.

Until the epizootic, horses seemed like an ideal energy source for Chicago. Their digestive systems quickly converted calories from hay and grain into forward movement. With the right rigging, horses could be used for various kinds of work: pulling streetcars and omnibuses, hauling freight, towing fire engines, drawing delivery wagons, carting loads of scrap or garbage, and carrying riders.

Horses could memorize routes and operate with little human intervention. They also provided the motors behind the characteristic clip (and clop) of the modern metropolis. Any urban energy source had to meet the demands of the fledgling city: a large population that lived farther from work than ever before, a booming market economy, as well as the rapid pace to which urbanites had become accustomed. With all their horses out sick, newspaper editors, merchants, and transit company owners embarked on a quest to meet these demands and identify a non-horse-powered source that could fit—both structurally and culturally—in an urban environment designed for horses.

The 1872 equine influenza epizootic profoundly disrupted the city and changed how horses fit within their niche. In ecology, a niche refers to the interaction between a particular species and its habitat. To survive, the animal responds to changes in the habitat and shapes the environment in return. Horses, through their work with and for humans, filled a particular energy niche in the nineteenth-century urban habitat.[4] Because the habitat was structured for horses and humans, new transportation technology like steam and electricity had to fit in a horse-size niche to succeed. By reconstructing the street-level human-animal city and tracking human responses to the epizootic, this essay offers a city-level analysis against the traditional hoof-to-tire narrative, which sees electric transportation and the automobile as replacing horsepower in a neat, straight line through the early twentieth century. Instead of steam and electricity supplanting horsepower, new energy systems at the turn of the century evolved upon physical and cultural foundations designed for and by animals: *for* them when humans designed roads and buildings around equine physiologies, and *by* them when equine intelligence and metabolic systems supported vast transportation networks and economic growth.[5] In Chicago, the success or failure of urban street technology depended not merely on human innovation but also on whether technology could function like an animal in an ecological niche defined by overextension and interdependence. Any new urban transportation technology, in other words, would have big horseshoes to fill.

Chicago's horses were central to the industrializing process, both in terms of defining infrastructure and technology and in shaping the new "modern" culture of a city. By viewing Chicago's epizootic experience through a cultural and industrial lens, we can better see how human dependence on animal labor shaped industrial landscapes and cultural practices. To read the epizootic as both built environment and culture, this chapter bridges two approaches to urban animals: on one side, the mechanical and labor histories of urban animals, and on the other, cultural studies of how human-animal interactions shaped human and animal systems.[6] Examining cities as built structures as well as cultural ones has important implications for urban animal history. This perspective overturns the assumption that humans are solely responsible for and have ultimate control over developments in industrial science and technology—especially those that happened around the turn of the century. Dethroning human control over the urban landscape helps us reject that false boundary between urban and natural environments and lets us better see the unintended technological and cultural consequences of our interconnected human-animal ecosystems.[7]

RECONSTRUCTING CHICAGO'S OUTBREAK

The 1872 equine influenza strain was especially virulent for two key reasons. First, the virus was contagious during both its first and secondary infection phases, so horses who seemed to be recovered could still shed enough of the virus to infect their stablemates.[8] Second, the virus had a rapid onset, making horses appear healthy one minute and sick the next.[9] Such rapid onset delayed identification of infectious horses and allowed the virus to spread as it silently incubated in visually healthy bodies.

The virus quietly entered Chicago on Sunday, October 20, 1872, when ten horses and their groom arrived at 612 West Jackson Street. The apparently healthy horses had been imported directly from Canada, where the virus had originated several weeks before. Two days later, one of the horses began to exhibit hallmark symptoms of the flu: a cough, runny nose, and lethargy. By Thursday, October 24, eight of the horses were sick with influenza. Attempting to spare the two remaining horses, managers transferred them to an Ashland Avenue streetcar stable located at 609 West Madison Street. Just forty-eight hours later, those two horses had infected twenty more.

Despite thorough newspaper coverage of the raging outbreaks in Buffalo, Rochester, and New York, barn managers, teamsters, carmen, veterinarians, and reporters remained skeptical that the virus would reach Chicago. The Chicago *Evening Mail* echoed the profound denial and suspicions of false news teeming in the city: "It is fortunate that horses can't read the morning papers, otherwise our Chicago equines would be scared to death today on learning that three months ago there were several cases of the 'Canada distemper' in one of the street railway stables. However as there is no evidence of the presence of the dreaded epidemic in this city now, they would not thank the morning papers for their unnecessary attempt to be sensational."[10]

Most stables declined preventive disinfecting measures; even those that started disinfecting regimes cast their lots with overconfidence and wishful thinking.[11] Even Dr. Rauch, the city's sanitary superintendent, remained doubtful, "convinced that there [was] not a single case of the 'Canada Consumption' in the city."[12] Nevertheless, newspaper coverage depicts a city on edge. Editors fretted over the city's most vulnerable stables, many of which kept horses in less than ideal conditions. Dr. Tuthill, the veterinarian responsible for the West Side Car Company's 320 horses, told the *Tribune* that he expected the influenza to arrive and do considerable damage to the city's barns. "Should the disease prove to be the dreaded epidemic, as he had no doubt it would, the dingy, dirty, ill-ventilated, character of the livery and boarding stables, omnibus stables, and auction stables together with the crowding which they were compelled to put up with, would produce a terrible mortality," reported the paper.[13] The city's many underground stables posed a particular challenge to ventilate and clean. Worried transit companies modified horses' work schedules to prevent exhaustion and decrease circulation through the streets.[14]

By Monday, October 28, Frank Parmelee's omnibus horses—among the finest cared for in the city—began to cough.[15] The following morning, sixty of Frank Parmelee's horses were sick. In Young's Omnibus Stables, "one could not hear his own voice for the coughing," according to the newspaper.[16] A few days later, more than forty barns had sick horses. The *Tribune* reported that more than three hundred cases had appeared in a twenty-four-hour period.[17] Reconstructing the timeline using stable reports published by the *Tribune* exposes that there were upwards of a thousand cases of the virus in the city just eight days after the equine patient zeros had arrived. Horses throughout the city appeared cold, dull, and "dumpish"; many had a cough and refused to eat.[18] Even the fire department horses were sick. Because fire horses (like firemen) were trained to assemble in formation in response to the alarm bell, fly through urban traffic pulling a heavy engine, and remain calm when they encountered a burning structure, this was an especially dangerous development.

The first few days of November saw a citywide shutdown. The Chicago

City Railroad Company started running its dummies—light steam locomotives disguised as railroad passenger cars—after midnight to make up for the missed daytime loads.[19] Field, Leiter & Co. sought an older form of motive power, hiring teams of oxen to make their dry-good deliveries through the city.[20] The horses of the Illinois and Michigan Canal were pulled from duty. So many post office horses were sick, mail had to be carried by hand.[21] Public transportation was at a standstill, and the city council held a special meeting to extend the use of dummy engines during daytime hours.[22] Prior to the epizootic, the expensive, accident-prone, and smoke-belching engines had been relegated to the suburbs or only approved for stockyard or nighttime snow removal.[23] Special approval for extended use of loud, smoky, and expensive dummy engines throughout the city reflected the urgency of the crisis. November 6—election day—was wet and bleak in weather and spirit. Rain and fog plagued the deserted city with mud and misery. The widespread unavailability of horses made voting laborious and bled the fanfare out of election day.[24]

November 7 marked a turning point in the outbreak: horses began to either recover or perish. A rash of fatality reports flooded in from the fire department, the US Express Co., and four other stable facilities.[25] Sick horses had been put to work too early, prompting Dr. Rauch to call upon the police powers of the Humane Act and begin forcibly removing sick horses from the street.[26] Sixty more horse deaths, including the fire chief's horse, were reported the next day.[27] Over the following days, Chicago's surviving horses quietly recovered, but it would be another week before they were back at work on a citywide scale.

In the end, according to the Chicago Board of Health, 1,150 horses died during the outbreak.[28] A Board of Health report calculated that for the year ending in March 1873, the city removed 2,873 dead horses. The following year, the city removed 1,812 dead horses, nearly a 40 percent decrease.[29] By accounting for the rising population of horses in a rapidly expanding city and remembering that the virus had a low fatality rate with proper rest and care, we can appreciate the mortal toll the nineteenth century's untenable equine energy system took on its horses.

HORSEPOWER IN THE CIRCULATORY CITY

In 1870s Chicago, horses provided the hauling power for the city's streets—its circulatory system—in great numbers.[30] Streetcar horses enabled the regular, rhythmic pulse of people and goods via timetables. Horsecars left intersections at intervals of one, four, and eight minutes.[31] Streetcar and omnibus horses carried people between home, work, church, and stores. Horses employed by railroad and omnibus companies filled in the gaps created by the rapidly expanding but decentralized overland transit system.[32] The city's estimated 25,000 horses crossed paths as they followed individualized routes dictated by different companies or loads.[33] Infected horses also moved through this intricate system, first carrying the virus along well-traveled streetcar lines. Located along the city's main streets, transit stables housed up to 350 horses in one building. Madison Street, which runs east to west, had at least eleven major transportation stables; State Street, which runs north to south, had at least six.[34] As these public transportation horses worked and rested on the crowded streets, they coughed, sneezed, snorted, and drank from water troughs, passing the virus on to livery stable horses hired out for the day as well as privately owned horses who also traveled city streets. Horses employed by railroad and baggage companies were likewise susceptible through the shared space of the street as they moved between depots. Many expected privately owned horses and carriages to make up some of the lost transportation, but private stables and expensive horses were swiftly infected by the virus thanks to the circulation of all kinds of horses through the city streets.[35] The very flexibility, efficiency, and sociability that made horses ideally suited to power the city made both the city and its animals especially vulnerable to influenza.

In Chicago's urban ecology, as in all horse-powered cities, humans perfecting the laboring animal body bred dependence on the energy it could create.[36] Equine bodies provide exceptionally efficient forward motion: their fixed spines minimize "energy-sapping vertical motion," their internal organs act like great pendulums gaining momentum as they swing in the thoracic sling, and horses' quick digestion allows them to quickly access energy from calories.[37] In this way, horses generate strong but enduring motive power. These features did not occur by happy accident, however. Humans engineered horses for size, temperament, and conformation through selective breeding; in doing so, humans created a form of *biotechnology*, an organic being for human use.[38] And when humans combined horses with inorganic technology like the harness, the horse collar, or the carriage, they created *cyborgs* or "complex system[s] of mechanical and biological characteristics integrated to perform certain kinds of work."[39] Over millennia, humans had shaped horses into a living technology designed for work on city streets; in return, the labor of these cyborg horses shaped human life in the city structurally and culturally.

Humans constructed urban built environments according to horses' bodies and needs. The particular traction requirements for hooves dictated road grading and paving materials; the turn radius of horses—lateral movement of their bodies naturally limited by their bone structure—dictated cities' street and block layout. Public water troughs for horses dotted the streetscape. The daily needs of different types of horses shaped the size, material, and architectural elements of building facades.[40] Horse-drawn streetcars, also known as horsecars, radically altered city boundaries by determining where individuals lived and what could occur within those borders. Streetcar fares reinforced human geosocial hierarchies; wealth increased the farther one lived from the city.[41] For a time around 1860, Chicagoans used horsecar tickets as a form of currency for small change.[42] The rigid structure of streetcar rails reorganized the lived experience of urban streets. Rather than the dynamic and chaotic spaces of conducting business, playing, and socializing that streets had been earlier in the nineteenth century, streets became spaces for conveying travelers *through* the city after the installation of rails.[43]

Such a connected and intelligent transportation system depended upon mutual understanding and emotional bonding between horses and humans. To work successfully with horses, humans mimic a naturally occurring horse hierarchy and herd behavior. In both wild and domesticated herds, horses enforce a distinct social order through dominance signals like pinned ears, biting, kicking, and squealing. The herd also engages in mutual grooming, coordinated nose-to-tail fly swatting, and play to bond members of the herd. Humans use brushes, whips, and verbal commands to mimic the dominance and bonding behaviors found in the herd and establish humans at the top of the pecking order.[44] Further, humans surgically geld (castrate) males to make them even more submissive and willing to bond with their handlers.[45]

The time Chicagoans spent grooming, naming, and learning individual horses' preferences, limits, and quirks ensured the smooth running of the city's transportation system.[46] Teamsters and drivers named their horses, often reflecting unique personality traits or physical attributes of the animal.[47] Names established emotional relationships between driver and horse, but names served practical purposes too. A growing city's streets were loud and chaotic: drivers' shouts and whistles mixed with cracking whips, streetcar bells, and the clatter of iron shoes on pavement. Because horses recognize their names and can be trained to respond to distinctive sounds, names enabled drivers to cut through the noise and communicate with their own horses.[48]

Driving horses in the city required specialized knowledge of more complex horse behavior systems. When deciding which horse to place in which harness position, drivers had to manage personalities and leadership roles. The dominant horse in a herd often changes in different contexts; the leader of the pasture or stable may be more submissive while working in harness. Successful drivers needed to first identify the working leader from the herd and harness that horse not at the front of the team but in the middle or at the wheel, so the horse could lead with more strength and balance.[49]

Horses also performed important intellectual labor in the city, and they were valued for what they knew and could learn about city navigation. Equine brains can memorize patterns, and many have a strong sense of direction.[50] Horses learned their routes, and milkmen and trash collectors trusted their horses to walk the route as they carried milk door-to-door or hauled garbage from the curb.[51] Horses' sense of direction enabled them to drive themselves home, delivering sleeping doctors safely after late-night house calls and returning exhausted streetcar operators back to the barn.[52] By learning their routes and executing them without the supervision of their drivers, horses became highly specialized laborers. Moreover, that drivers trusted their horses with carts full of valuable goods or drivers' own sleeping bodies reflects a high degree of emotional bonding and familiarity between individual humans and individual horses. Relying on innate equine bonding and intelligence, these complex human-animal systems enabled Chicago's delivery and transit systems to function efficiently and profitably.

But the meticulous and interconnected ecology of human-equine relationships and labor that made the city function also made it fragile. The epizootic revealed the degree to which Chicago's public and private spheres were interrelated and depended on healthy animal bodies to run. The initial loss of transportation produced a ripple effect over the rest of the city. The epizootic affected seemingly unrelated departments, institutions, and cultural establishments because these municipal functions relied on horses. Without horses to transmit prospective buyers to plots on the fringes of town, suburban real estate sales ceased.[53] Theaters, located away from residential areas in a neighborhood people called the "Burnt District," sat empty without horses to deliver patrons to their doors.[54] The city coroner had no way to transport human remains; the city asylum could not move patients.[55] Horses, in other words, were the unexpected link between Chicago's opera houses and its morgue.

Without horses to move people in the fall of 1872, the local and national elections suffered. For one *Tribune* contributor, the epizootic undermined the momentum of the presidential campaign just before votes were to be cast. Thousands of sick horses became the top priority almost overnight, drawing public discourse and enthusiasm away from politics. Rather than "For whom will you vote?" Chicagoans asked, "How is your horse?"[56] At its height on election day, the epizootic limited available transportation to Chicago polls. Days earlier, the *Tribune* described this worrisome side effect: "Wherever the disease prevails in the rural districts it will be impossible for the farmers to get to the polls. In the cities the vote will probably be reduced, for that considerable number of voters who always depend upon wagons, and won't or can't vote on foot, will be wanting."[57] Voter participation was vital to Chicago's ability to function as a municipality and ensured its place within the larger, national political context. That horses could risk voter participation and turnout made horse-powered transportation a threat to the city's political future.

The city's markets also felt the squeeze of absent horse-powered transportation. An editorial contributor in the *Tribune* opined that the epizootic exposed "[with] how slender a thread our business is upheld."[58] Farmers decreased their imports to the city for fear of infecting their horses.[59] With so many horses off their feed, the price of oats stagnated unexpectedly.[60] Until firms acquired teams of oxen, coal, wholesale grocery, and dry good deliveries in the city stopped. When the horses were pulled from the Illinois and Michigan Canal, stone, coal, and grain shipments to the city faltered.[61] Without horses to drive cattle to slaughter in the stockyards, the city's meat output decreased.[62]

Not only was the city not moving or trading as it should, but the city did not *feel* as it should. Without horses, Chicago was not the bustling city to which Chicagoans had grown accustomed. As company owners and barn managers pulled horses from their routes and service, the city streets grew quiet and still. The clang of iron horseshoes on pavement, the hollers of drivers, and the snorts, neighs, and nickers of horses were abruptly absent. The silence was "painful," and many wrote of the "depression" that seeped into the city psyche without horses in the streets.[63] The return of the sensory experience of the city served as one of the earliest indicators that the crisis was abating.

The newspaper rejoiced when "the tingle of the street-car bell was again heard, and the omnibus lines resumed their usual routes."[64] Hallmarks of urban life, this noise and speed—both generated from horsepower—defined the lived experience of modern, industrializing Chicago.

In 1872, public transportation systems, cultural institutions, political engagement, and market exchange helped define Chicago as a modern metropolis. The swift and noisy circulation of people and goods through that environment characterized its sensory experience. As the epizootic made painfully clear, horses were at the center of urban life—but horses were no longer suitable to fill this crucial niche.

OX-POWERED CHICAGO

That a disease could so swiftly and so pervasively sicken the city's energy source and halt urban progress was troubling. Realizing they had considerably less control over their city than they expected, humans began a search for more sustainable energy sources to power Chicago's streets. This new power source needed to fill a highly specialized niche that had structurally and culturally evolved around animal bodies. To replace horsepower, it needed to be flexible, efficient, fast, and able to be implemented on a citywide scale. But beyond practicality, this power source needed to fit within human expectations of "modern" urban culture.

Initially, Chicago fell back on other kinds of animal power. Substituting species rather than machinery was easier due to the city's animal-designed structure and the duration of the outbreak. Oxen especially, but also donkeys, goats, and even dogs, were called upon to provide motive power.[65] Because Chicago's streets and transportation networks had been built for horses, the other species impressed into service were not always ideal fits. Many of the new animals were inexperienced. A young donkey who "had scarcely outgrown flirting, gay clothes, and unnaturally strong cigars, and [who] was not as much at home on the street as he should have been" was spotted pulling a man and his cart.[66] Dr. Charles W. Zaremba of the Illinois Humane Society implored drivers "to load carts proportionately and be knowledgeable about driving and yoking" as the oxen imported from the hinterlands into the city were young and far less experienced.[67] For Chicago, these other species, but particularly oxen, had the potential to fill horses' place mechanically but not culturally.

Such a monospecific city could not replace its laboring animal population with anything other than horses.

Oxen arrived in the city as hardworking and economical relics from a traditional American yeoman past, and their sudden appearance in the city represented a temporal and spatial break and the return of a pastoral ideal.[68] The initial prime mover in colonial America, castrated male cattle were strong and powerful but slow. Their steady, strong bodies came with tolerant, easily trained brains. Oxen typically "were used wherever people needed cheap power but not speed, in regions with subsistence agriculture, little commerce, and poor roads, and for jobs requiring very grueling draft work."[69] New Englanders equated oxen—popular in the difficult-to-farm region—with wealth and success, but to the rest of America, horses were the animal symbol of upward mobility and social prowess.[70] When oxen plodded into Chicago, the *Tribune* rejoiced: "We shall now have a chance to return to first principles in trade, the ox being the real pioneer of domestic animals, in clearing the forests and tilling the soil."[71] As more firms felt the pressure to resume business, there arose "ox markets all through the business portion of the city, and about them gathered merchants, ox-drivers, and curiosity-seekers."[72] Fascinated, Chicagoans welcomed oxen into the urban fold and shaped the species and themselves in the process.

Oxen remained a short-term solution to the epizootic, but the city adapted to ox power far more quickly and effectively than it did to inorganic alternatives like steam. Like horses, oxen bonded with drivers. But unlike horses, oxen worked in yokes and responded to different verbal commands.[73] The city's shortage of skilled ox handlers meant drivers—many of whom were clerks coerced into driving service—often had to form a relationship with their animals and learn new driving techniques on the spot. A *Tribune* reporter witnessed clerks struggling to communicate with their team of oxen: "They first attached tug-ropes from the yoke to the whippletrees, only to encounter the most uncompromising objections from the oxen, whose feelings found vent in kicks slow but disagreeable. In the midst of their quandary came a timely suggestion from a looker-on who was evidently well up in ox-gear. From him they learned that nothing but a chain of suitable length was necessary, and, when this had been procured, they went on their way rejoicing, at the rate of two miles per hour."[74] The oxen's resistance clearly identified the clerk's hitching errors, and with a modification of equipment, the satisfied team moved forward together. In these moments, the animals trained the humans.

After trading the hinterlands for the big city, the first oxen met inexperienced drivers who lacked ox-sense.[75] In an unfamiliar world of tall buildings and bustling streets, the oxen were overdriven to exhaustion, their soft feet damaged without protective shoes, and their necks choked from improper yoking.[76] However, drivers' skills and ox-sense improved over a matter of days, reflecting human familiarity and experience with animal relationships. A history of human-animal work meant conditions for the oxen swiftly improved. "Many [oxen], too, have discarded the rude and cumbersome yoke, and are now dressed in fine leather harness, and lines are substituted for the ox-goad," proclaimed the *Tribune*. "These many improvements will certainly make the ox more comfortable, and he will believe that Chicago is rather a humane place after all."[77] This passage reflects the city's flexibility in adapting to ox-power, but it also suggests that urban work and the urban environment shaped oxen in return.

Exchanging wooden yokes for leather harnesses represented a transformation of the agricultural animal. Through their work in the city, oxen achieved the trappings of a more civilized, more urbane, more *modern* species. By compelling oxen and humans to work together, the epizootic impelled Chicago's company managers, drivers, clerks, public witnesses, and even *Tribune* readers to reinscribe the cultural meaning of oxen. So many oxen were brought in to make city deliveries that the newspaper credited the animals with saving the city's business economy.[78] These accolades temporarily elevated oxen from pastoral curiosities to noble stewards of the urban market. Nevertheless, the oxen were slow and could not match the fast-paced thrill of urban life set by horses. Through their temporary work in the city's streets, oxen gained higher cultural and social value but ultimately returned to their farming roles in the hinterlands once the city's horses were well enough to resume their positions.

Their structural footprints designed for animal power, transportation companies and businesses swiftly adapted to slow draft animals and made space for the improvised use of dogs and donkeys. A history of human-horse partnerships made other animals the first choice to fill the niche left by horses. Through this process, experiences with new animal species reframed public perceptions of those animals' symbolic values and meanings. These animal alternatives, however, were too slow and inefficient for long-term adoption and were particularly unsuited to power vital public transportation systems. Rather, Chicago's streetcar lines looked to steam, believing steam power could assume the speed, efficiency, and flexibility demanded by a city ecology built around horses.

THE LIMITS OF STEAM

In addition to oxen imports, the epizootic accelerated efforts to adopt new transportation technology. With horses suddenly absent from their place on Chicago's pavements and rails, debates over the merits and potential of steam bloomed. Newspaper editors and investors saw the epizootic as an opportunity to use the city's deserted streets as steam-powered transit's proving ground and a means to develop better theory through practical implementation.[79] For streetcar company owners suffering enormous profit loss and Chicagoans tired of walking, steam appeared as a superior transportation technology, and the epizootic had offered up steam as the way out of major circulatory system problems.

Steam-powered streetcars promised to draw passengers "more economically, safely, swiftly, cheaply, and conveniently" than horses ever could and relieve street congestion problems at the same time.[80] On a grander scale, steam locomotives offered to completely do away with the horse-powered transfer system railroad companies and passengers needed to overcome track gauge differences. In the *Tribune*, proponents pitched a centralized railroad depot supported by a network of radiating steam railways. Located near the stockyards, "a gigantic transfer freight house" would take in the city's freight, "transferred by switches, without change of bulk or hauling into the city."[81] This grandiose plan (which the author questionably asserted could be *mostly* enacted in ten days) would eliminate the need for heavy-transfer horses entirely.[82]

During the epizootic, the steam engines brought in to keep Chicago's transportation system moving proved to be capricious power sources. Steam was not new technology in 1872, but it was new to urban transit. By the 1870s, steam power was the first choice for long-distance travel and factory machinery. Over shorter distances—the very distances over which horses excelled—steam was inflexible, unreliable, and dangerous. Steam machinery was likewise too fast, too smoky, and too prone to exploding for many city dwellers.[83] In Chicago, successful use of steam dummies on the North Side line was celebrated. But the South Side line belonging to the Chicago City Railway Company met with "calamity after calamity" when accidents damaged the engines and made them useless for further runs.[84] One engineer overfilled an engine, and the boiling water exploded in his face.[85] Further, companies could not implement steam systems quickly enough. Whereas oxen required some modification of behavior, harnessing equipment, and special iron shoes to successfully work on the street, steam necessitated new urban infrastructure. Railway companies could not build steam-powered boilers, engines, and streetcars overnight.[86] Finally, because pavement and intersections had been built for horses, not all steam engines could navigate sharp curves, and the public did not trust the engines' ability to brake in traffic.[87] Chicago's businesses had adapted to oxen virtually overnight, but transit companies struggled with steam.

Another objection to steam-powered streetcars in Chicago was that the loud apparatuses would frighten horses working on the street.[88] Shying and balking horses shaped technological design when manufacturers believed horses feared the shape of the engine (rather than its shrill sounds and belching smoke) and built steam dummies to look like passenger cars. Such fears also confirmed that horses were the rightful inhabitants of the streets and steam engines were the interlopers. In a letter to the *Tribune* editor, "H," self-identified as a horse owner with no interest in railroad companies, postulated: "Which has the best, first right to the use of our streets: the dummy or the horse? Would the public interest be best served by yielding that right to the dummy or the horse?"[89] He answers his question with the assertion that steam could be successful only through the education of horses. Horses, rather than steam, remained the flexible, intelligent technology. The author argued that a city ordinance "providing against the use of uneducated horses, by proper provision for their inspection and training, at a minimum fee" would allow Chicago to "move in a lead which will be followed by the world."[90] For "H," horses' capacity to learn and municipal regulation were the keys to propelling Chicago into a technologically superior future. Chicago's primacy as a metropolis was at stake, and inorganic energy like steam was the ticket to modern technological progress. Development of this technology depended upon human-animal relationships and the coevolution of steam machinery and equine learning.

Other *Tribune* accounts echoed this process. One contributor pointed out that horses had grown accustomed to steam locomotives with their "noise, and smoke, and puffing, apparently [seeing] sufficient resemblance in the railway-engine to themselves to understand it." The problem with dummy engines, according to the article, was that the disguised motors *concealed* the mechanism for motion and confused the rational horse. Without a clear explanation for where the power came from, horses reacted with fear and suspicion.[91] "B" believed the quandary could be solved by hitching either a real or a mock horse to the front of the dummy engine. "B" writes, "According to my view, the mental state of a horse upon the approach of the dummy is this: Astonishment, ending in fear, not at the dummy nor its train, but at the phenomenon, for the first time presented to his vision, of matter in automatic motion. Deceive him by showing him a natural cause,—a fellow horse,—and his trepidation ceases."[92] It was simple: provide a logical explanation for movement by modifying the machine, and the horse will settle. To properly take the place of the horse in Chicago's streets, the steam engine had to not only function like a horse but also look like one.

Steam proved capable of upholding the bustle of urban life set by horsepower but only at the expense of being unpredictable and rigid. Horses' natural behavioral responses to fear—startling, shying, bolting, rearing—and their capacity to accept frightening stimuli altered the perception and implementation of steam power in Chicago. As they did with oxen, humans had to adapt their own behavior and inorganic technology to suit the needs of animals' bodies and minds. In this way, urban street transportation systems developed in both literal and conceptual animal spaces and evolved through adaptations designed to meet particular animals' needs. Steam had to match the pace, expectations, and built environment set by horses; its inability to do so ultimately undermines a linear understanding of the evolution of urban transportation. Rather than replacing horsepower outright, steam, and eventually electricity, coevolved with it.

RECONSIDERING THE URBAN HORSE

During the epizootic, horses began to outgrow their cultural niche in addition to their technological one. While humans searched for more sustainable energy alternatives to horses, business owners, barn managers, and private citizens realized that horses underpinned Chicago's success and urban growth. But with the city shut down, stakeholders also had to provide more humane and humanlike medical care to nurse those vital laborers back to health. Such human(e) treatments pushed horses farther away from being considered "living machines" and toward recognition as empathetic, valuable beings who warranted better living and working conditions for their service.

The outbreak prompted public recognition that horses propelled the city structurally and culturally to great heights. Such assistance was all the more valuable given the recent conflagration. "[The horse] helps to rebuild the city, conveying the material to the buildings and elevating it to the topmost story," wrote a *Tribune* editor.[93] Verticality, a hallmark of the modern, industrializing city, came down to horses. Moreover, Chicago became aware that horses buttressed the city's economic and cultural spheres. "Now that the retail merchants find their stores empty, they will be inclined to place a higher estimate upon the lines of street-cars and stages which the horses propel," declared the *Tribune*. The loss of the urban experience increased the value of equine life and labor, and Chicago's forward and upward momentum was no longer the exclusive domain of its human population. If humans were no longer in control of urban success, the very animals that made the city strong also made it vulnerable.

During the epizootic, horses' importance to the urban ecosystem came to the forefront, and their medical care challenged expectations about how they should be treated. The crisis knit together human and animal health. Without widespread, formalized training for professional veterinarians, the city fell back on its experience with human disease. Municipal management of the outbreak harnessed past experiences with human epidemics and applied public health principles to combat equine disease. Even before the virus had infected thousands of the city's horses, the *Tribune* drew parallels between the smallpox epidemic of 1871 and the current equine crisis.[94] Under sanitary superintendent Rauch's supervision, the city and stable managers used standard public health measures of quarantining and isolation to prevent the spread of the virus. The Illinois Humane Society shut off public horse water troughs and provided individual pails instead.[95] Disposal of infected carcasses was arranged with the Union Rendering Company and the day scavenger contractor, and when Maggie, a mare belonging to A. J. Wright's Livery Stable, died of complications, her body was dissected to better understand the disease.[96] The overlapping spheres of animal and human health were not new for the city, as cowpox pustules were central to the inoculation of thousands of Chicagoans against smallpox.[97] During the epizootic, instead of animal health being used to improve human health, Chicago's municipal government brought human healthcare to horses. By managing equine influenza as they would human influenza—using quarantine, counting cases, preventing cross-contamination, and organizing the disposal of remains—city officials extended municipal oversight to the animal city, further bringing animals alongside humans under the public health purview. Healthy animals, in this way, became paramount to a healthy city.

As on the citywide scale, treatments in barns mirrored medical care for humans. Sick horses transported teamsters, drivers, hostlers, grooms, and even clerks into caregiving roles. So long as horses were given adequate medical care and rest, the influenza virus had a low mortality rate. Veterinarians, public health officials, and well-respected horsepersons championed treating sick horses like sick humans: the stable environment should be kept clean, regularly disinfected, and ventilated.[98] Horses should be kept warm and fed good hay, apples and carrots, and oatmeal gruel.[99] The *Tribune* reported that Frank Parmelee's "horses have taken to drinking slippery elm like sick children."[100] Further suggestions for humanlike treatment included hay tea and an improvised steam bath, created by placing a heated brick into a bucket of water and placing a blanket over the horse's head, to soothe damaged respiratory tracts.[101] The bonds between horses and their handlers even formed part of the treatment protocol. Dr. Paaren, veterinary editor of the *Prairie Farmer*, recommended that strangers be "forbidden to enter the stable; none must approach the animal but the man it is accustomed to, and he must move as quietly as possible."[102] The close parallel between treating sick horses and treating sick humans helped support a larger movement of humans learning to see animals, and particularly horses, as kindred spirits, intelligent creatures with the capacity to suffer.

An editor in the *Inter-Ocean* summarized Chicagoans' changing relationship with the horse: "We never knew till now how much we owed to that noble animal, the horse. To be sure we always admired him, loved him, and taught our children to regard him as the noblest of all quadrupeds, the paragon of animals. But we did not quite realize the extent to which we were dependent on him till he was taken from us."[103] Human empathy and sympathy for horses existed prior to the outbreak. But a mass reckoning with the intertwined fates of urban horses and urban people? That was new for Chicago.

The profound shifts sparked by the epizootic helped ignite humane advocacy that had been simmering since midcentury and altered the cultural expectations and implications of equine labor in the city.[104] The *Tribune* denounced caretakers who failed to provide empathy and companionship, publicly reinforcing the new horse-human relationship. As with human medicine in the late nineteenth century, not all prescribed treatments were salubrious. "Aconite, cayenne pepper, alcohol, chloride of lime, mustard, carbolic acid, lobelia, belladonna, spirits of nitre, Sulphur liniment, burnt hops, burnt brimstone, bromo-chloralum, condition powders" were among the more toxic remedies used at stables across Chicago. The *Tribune* cautioned against use of these tonics. At the same time, the paper shamed companies that chose profit over care by putting horses back to work too soon and condemned sham veterinarians peddling snake oil cures.[105] Newspaper coverage helped enforce the new position that the loyal creatures who helped build and then rebuild Chicago deserved human and humane treatment.

The epizootic further eroded the boundary between human and horse by altering labor patterns in barns and in the city streets. When oxen or steam could not make up for the loss of horsepower, humans had to assume that work, overturning the expectations of what jobs were for humans and what jobs were for animals. With their horses on stall rest, Chicago's post office had to establish a "wheelbarrow express" to circulate hundreds of letters and packages through the city.[106] Carpenters, plumbers, milkmen, as well as baggage and express wagon drivers found themselves running entirely on manpower.[107] When fire department horses began to fall sick, plans were made to ensure that ropes would be available for men to pull the engines.[108] The largest labor reversal of all was that to get anywhere, most Chicagoans had to walk. At once, the epizootic elevated the social position of horses and temporarily reversed the roles of human transportation in the city.

MOVING FORWARD AND LOOKING BACK

The fall of 1872 was a reckoning for Chicagoans. The *Tribune* proclaimed, "The horse is somebody. . . . Everywhere and among all classes . . . the result of the present visitation must be a practical revaluation of the horse which will tend to give him his true place in civilization, and, it is to be hoped, ameliorate his condition in life."[109] Yet urban dependence on horse labor prevailed for another half century. Neither other animals nor steam could fill the urban ecological niche set by horses. With time, however, popularization of bicycles and electric trolleys came close.

For the three decades after the epizootic, a growing urban horse population supported Chicago's continued expansion. By 1900, more than 73,000 horses lived in the city.[110] At the same time, the popularity of the bicycle and expansion of electric trolley lines created a more stable urban ecosystem by introducing mode diversity into the city's transportation network. In an 1897 article for the *Chicago Inter Ocean*, a contributor even quipped, "Well, the poor horse could get over the epizootic distemper, but he could not recover from the bicycle and the trolley."[111] According to one *Chicago Chronicle* editor looking back on the epizootic from 1896, this new energy system had taken root in the 1872 outbreak: "[The epizootic] greatly stimulated street railway building, and no doubt set inventors to the thinking out of devices to supplement or take the place of draft animals. Anyhow, street cars are now everywhere, inventions have been made, and now, in existing conditions, the coming in epizootic form of any disease of horses could not work anything like the vast inconvenience and loss that were occasioned by the strange affliction in the former time."[112] In a sense, the decline of the urban horse and the subsequent urban transit technology shift had begun in the early 1870s, several decades before the use and population of urban horses had even peaked.[113]

The epizootic likewise loomed large in public memory. Those few weeks in October and November 1872 remained a cautionary tale about the city's vulnerability to disease and the interconnected fates of humans and animals. When influenza threatened the human residents of Chicago just before the new year in 1889, a newspaper editorial used the horse flu to admonish the city's doubters. The article calls upon memories of the epizootic's biblical appearance and devastation: "Those who remember the complete prostration of business in this city in 1872 by the epizootic, which came upon us like a thief in the night and hung around like one of the plagues of Egypt, will hesitate before pronouncing such a human epidemic impossible."[114] Editors recognized the worrisome pattern of news reports pouring in from other countries and cities and the systemic consequences that had come from a city refusing to take a virus's potential seriously.[115]

As the urban horse population grew, so did humane advocacy in Chicago. The epizootic had changed expectations about the value and emotive power of the industrial workhorse, and those sentiments provided fodder for advocacy well past the turn of the century. For humane advocate and prominent Chicagoan Hortensia M. Black, the epizootic had been a watershed moment in revealing just how much the city owed to and depended upon its equine residents. In an interview with the *Chicago Tribune* in 1899, she argued, "Why or how we may not be able to say, but human and animal interests are inextricably bound together. It was estimated that the 'epizootic' affliction of horses . . . cost Chicago $1,000,000 daily loss in trade while it lasted, and gave a lesson on the value of the horse in ways that bicycles can never compete with." She continued, "Leprosy disappears when the cruel sport of fishing ceases. Tapeworm is the result of slaughtering cattle; trichinosis of the cruelty to hogs. In short, all life is filled with revenges of cruelty to animals and of rewards of kindness."[116] From Black's perspective, the epizootic was an intended consequence—and Chicago's just deserts—for its rampant exploitation of urban horses.

In examining Chicago's human-animal ecology during the 1872 epizootic, we realize that "modern" transportation technology in cities was not an exclusively human endeavor and that it emerged out of an overextended and untenable equine labor system. By removing horses—whose bodies provided the energy literally and figuratively driving urban life—the epizootic untangles an often obscured coevolution of animals, technologies, and urban culture. At once, the epizootic was a revealer of the precariousness of Chicago's energy system and an agent of technological change. When the epizootic revealed the equine city to be fragile, people sought new technology as a way out of the unsustainable ecology they had created. In the process, horses created the footprints and expectations of urban transit and an urban culture that lingers through the present.

NOTES

1. Clay McShane, "Gelded Age Boston," *New England Quarterly* 74, no. 2 (2001): 279–80.

2. "The Horse Disease: This Community Brought to Realize the Usefulness of the Equine Species," *Chicago Daily Tribune*, November 5, 1872.

3. Sean Kheraj, "The Great Epizootic of 1872–1873: Networks of Animal Disease in North American Urban Environments," *Environmental History* 23 (2019): 498.

4. Ann Norton Greene, *Horses at Work: Harnessing Power in Industrial America* (Cambridge, MA: Harvard University Press, 2009), 6. See also Clay McShane and Joel Tarr, *The Horse in the City: Living Machines in the Nineteenth Century* (Baltimore: Johns Hopkins University Press, 2007), 1.

5. Greene, *Horses at Work*, 9. Greene's research provides a thorough macro-level examination of this coevolution process. This chapter uses the 1872 epizootic to

build upon her substantial work, offering a city-level case study to offer a clear picture of how these changing relationships among humans, animals, technology, and the environment were bound to and affected one another on a micro level.

6. For the former, see McShane and Tarr, *The Horse in the City*; and Greene, *Horses at Work*. For the latter, see Kathleen Kete, *The Beast in the Boudoir: Petkeeping in Nineteenth-Century Paris* (Berkeley: University of California Press, 1994); Susan Nance, *Entertaining Elephants: Animal Agency and the Business of the American Circus* (Baltimore: Johns Hopkins University Press, 2013); Harriet Ritvo, *The Animal Estate: The English and Other Creatures in the Victorian Age* (Cambridge, MA: Harvard University Press, 1987). Frederick L. Brown's *The City Is More than Human: An Animal History of Seattle* (Seattle: University of Washington Press, 2016) offers an exemplary example of how such a bridge provides an in-depth look at how various animals have helped shape a particular city. His argument that the presence of animals becomes a distinguishing factor in what is or is not considered a *proper* city is particularly useful in that it reveals animals as a defining feature of urban life.

7. See William Cronon, "The Trouble with Wilderness: Or, Getting Back to the Wrong Nature," *Environmental History* 1, no. 1 (1996): 7–28, https://doi.org/10.2307/3985059; Catherine McNeur, *Taming Manhattan: Environmental Battles in the Antebellum City* (Cambridge, MA: Harvard University Press, 2014).

8. Jeffrey Michael Flanagan, "On the Backs of Horses: The Great Epizootic of 1872," MA thesis, College of William and Mary, 2011, 8, http://search.proquest.com/docview/1964260200/citation/5DAB487D7A48400DPQ/1.

9. James P. McClure, "The Epizootic of 1872: Horses and Disease in a Nation in Motion," *New York History* 79, no. 1 (1998), 7; "The Horse Disease: Progress of the Pestilence in This City," *Chicago Daily Tribune*, November 1, 1872.

10. *Chicago Evening Mail*, October 24, 1872, 1.

11. "Look to Your Horses: The 'Canada Disease' Undoubtedly in Chicago," *Chicago Daily Tribune*, October 24, 1872; "The Horse Disease: Visit of Inspection to the Principal Stables of Chicago," *Chicago Daily Tribune*, October 25, 1872.

12. "The Horse Disease: Visit of Inspection to the Principal Stables of Chicago," *Chicago Daily Tribune*, October 25, 1872.

13. "Look to Your Horses," *Chicago Daily Tribune*, October 24, 1872.

14. "The Horse Disease," *Chicago Daily Tribune*, October 25, 1872.

15. "The Horse-Disease: Progress of the Pestilence in This City," *Chicago Daily Tribune*, November 1, 1872.

16. "The Horse-Disease: Its Presence Made Manifest in Our Streets Yesterday," *Chicago Daily Tribune*, November 2, 1872.

17. "The Horse Disease: Undoubted Evidence of Its Presence in Chicago," *Chicago Daily Tribune*, October 31, 1872.

18. "The Horse-Disease," *Chicago Daily Tribune*, November 1, 1872.

19. "The Horse-Disease," *Chicago Daily Tribune*, November 1, 1872.

20. "The Horse Disease," *Chicago Daily Tribune*, November 2, 1872.

21. "The Horse Disease," *Chicago Daily Tribune*, November 5, 1872.

22. "The Horse Disease," *Chicago Daily Tribune*, November 5, 1872.

23. See McShane and Tarr, *Horse in the City*, 5; Greg Borzo, *Chicago Cable Cars* (Mount Pleasant, SC: Arcadia, 2012), 233. An 1870 ordinance prohibited the Chicago City Railway Company from running its dummy on State Street but permitted its Stock Yard line from the city limits to 31st Street. See also *Laws and Ordinances Governing the City of Chicago* (Chicago: Bulletin Printing, 1873), 224. An 1871 City of Chicago ordinance gave the Chicago City Railway Company permission to run dummies to clear snow on its tracks between midnight and 5 a.m.

24. "The Election: The Result in Chicago Yesterday," *Chicago Daily Tribune*, November 6, 1872.

25. "Our Equine Patients: Danger Incurred by Setting the Horses at Work Too Soon," *Chicago Daily Tribune*, November 7, 1872.

26. "Our Equine Patients."

27. "The Horses: No Particular Change in the Condition of Our Equine Patients," *Chicago Daily Tribune*, November 8, 1872.

28. *Report of the Board of Health of the City of Chicago for the Years 1870, 1871, 1872 and 1873* (Chicago: The Board, 1874), 45.

29. *Report of the Board of Health of the City of Chicago*, 159, 169.

30. McShane, "Gelded Age Boston," 279–80.

31. Alfred Theodore Andreas, *History of Chicago: From 1857 until the Fire of 1871* (Chicago: A. T. Andreas, 1886), 121. This streetcar timetable comes from early 1871.

32. Greene, *Horses at Work*, 44, 75. At the time, tracks lacked a standardized track gauge, and each railroad line had its own depot on the outskirts of the city. Train function required horses to transfer passengers and baggage between depots, hotels, and coach stations.

33. Borzo, *Chicago Cable Cars*, 65.

34. Jennifer Marks and Jay Bowen, "Chicago Epizoozy 1872," https://jennmarks.github.io/Chicago_Animal_History/. This map uses GIS software to cross-reference historical maps, newspapers, and city directories to chart the spread of the equine influenza virus through Chicago.

35. "An Embargo on Trade and Travel," *Chicago Daily Tribune*, November 3, 1872.

36. Sean Kheraj ("The Great Epizootic," 495) calls this process "biotic homogenization."

37. Greene, *Horses at Work*, 17–21.

38. Greene, *Horses at Work*, 4.

39. McShane, "Gelded Age Boston," 292.

40. Greene, *Horses at Work*, 18, 45–46; Sherry Olson, "The Urban Horse and the Shaping of Montreal, 1840–1914," in *Animal Metropolis: Histories of Human-Animal Relations in Urban Canada*, ed. Joanna Dean, Darcy Ingram, and Christabelle Sethna (Calgary: University of Calgary Press, 2017), 70–80.

41. McShane, "Gelded Age Boston," 280. See also Sam Bass Warner, *Streetcar Suburbs: The Process of Growth in Boston, 1870–1900*, 2d ed. (Cambridge, MA: Harvard University Press, 2009).

42. Borzo, *Chicago Cable Cars*, 63.

43. Greene, *Horses at Work*, 185.

44. Greene, *Horses at Work*, 21–22.

45. McShane, "Gelded Age Boston," 283.

46. This is contrary to Hilary J. Sweeney's assertion that there was "little evidence of attachment to [horses] as pets or any type of social beings until late in the century" and her oversimplification that "if a horse was injured, too old to work, or simply worn out, it was often shot and the carcass sold to the tannery." Hilary J. Sweeney, "Pasture to Pavement: Working Class Irish and Urban Workhorses in Nineteenth Century New York City," *American Journal of Irish Studies* 11 (2014): 141.

47. McShane, "Gelded Age Boston," 293.

48. McShane, "Gelded Age Boston," 293.

49. Greene, *Horses at Work*, 26.

50. Greene, *Horses at Work*, 22.

51. Greene, *Horses at Work*, 22.

52. Greene, *Horses at Work*, 22; McShane, "Gelded Age Boston," 292.

53. "Real Estate: The Horse-Disease Interferes with Transactions," *Chicago Daily Tribune*, November 10, 1872.

54. "Amusement Review," *Chicago Daily Tribune*, November 10, 1872.

55. "Miscellaneous City News," *Chicago Daily Tribune*, November 10, 1872; "The County Commissioners: Transaction of a Variety of Business at Yesterday's Meeting," *Chicago Daily Tribune*, November 5, 1872.

56. "Political Influence of the Horse," *Chicago Daily Tribune*, November 3, 1872.

57. "Political Influence of the Horse."

58. "The Horse-Disease," *Chicago Daily Tribune*, November 1, 1872.

59. "Money and Commerce," *Chicago Daily Tribune*, November 4, 1872.

60. "Money and Commerce," *Chicago Daily Tribune*, November 4, 1872. That stables reported large numbers of horses refusing to eat, and the decrease in energy expenditure without their work in the street helps explain the low demand for oats.

61. "Money and Commerce," *Chicago Daily Tribune*, November 4, 1872.

62. "Money and Commerce," *Chicago Daily Tribune*, November 7, 1872.

63. "The Horse Disease," *Chicago Daily Tribune*, November 5, 1872; "Our Equine Patients: Chicago People Growing Weary of Walking," *Chicago Daily Tribune*, November 10, 1872.

64. "Convalescent: Our Equine Patients Recovering Rapidly," *Chicago Daily Tribune*, November 13, 1872.

65. "The Horse Disease: Another Day of Deserted Streets and Unhappy Pedestrians," *Chicago Daily Tribune*, November 6, 1872.

66. "The Horses," *Chicago Daily Tribune*, November 8, 1872.

67. "A Week without Horses: How the Epizootic Affected the People of Chicago," *Chicago Daily Tribune*, November 10, 1872.

68. Greene, *Horses at Work*, 10–11.

69. Greene, *Horses at Work*, 27–28.

70. Greene, *Horses at Work*, 29.

71. "Money and Commerce," *Chicago Daily Tribune*, November 4, 1872.

72. "The Horse Disease," *Chicago Daily Tribune*, November 6, 1872.

73. "The Horse Disease: The Streets Deserted and Business Brought to a Stand-Still," *Chicago Daily Tribune*, November 3, 1872.

74. "The Horse Disease," *Chicago Daily Tribune*, November 6, 1872.

75. C. Walsh & Son brought oxen in from Kenosha, Wisconsin, to deliver their mail. "The Horse Disease: Postal Facilities: The Mails," *Chicago Inter-Ocean*, November 4, 1872.

76. "The Horse Disease," *Chicago Daily Tribune*, November 6, 1872; "Our Equine Patients: Danger Incurred by Setting the Horses at Work Too Soon," *Chicago Daily Tribune*, November 7, 1872.

77. "The Horses," *Chicago Daily Tribune*, November 8, 1872.

78. "The Horses," *Chicago Daily Tribune*, November 8, 1872.

79. "The Horse-Disease," *Chicago Daily Tribune*, November 1, 1872; *Chicago Inter-Ocean*, November 5, 1872.

80. "Street Cars by Steam," *Chicago Daily Tribune*, November 13, 1872.

81. "[Our Rail System]," *Chicago Daily Tribune*, November 4, 1872.

82. "[Our Rail System]," *Chicago Daily Tribune*, November 4, 1872.

83. McShane, "Gelded Age Boston," 275–76.

84. "The Horse Disease," *Chicago Daily Tribune*, November 6, 1872.

85. "Our Equine Patients: Chicago People Growing Weary of Walking," *Chicago Daily Tribune*, November 9, 1872.

86. McClure, "The Epizootic of 1872," 19.

87. McClure, "The Epizootic of 1872," 19.

88. McClure, "The Epizootic of 1872," 19.

89. "Steam vs. Horses," *Chicago Daily Tribune*, November 9, 1872.

90. "Steam vs. Horses," *Chicago Daily Tribune*, November 9, 1872.

91. "Street Cars by Steam," *Chicago Daily Tribune*, November 13, 1872.

92. "Dummies and Horses," *Chicago Daily Tribune*, November 14, 1872.

93. "The Horse Disease," *Chicago Daily Tribune*, November 5, 1872.

94. "The Horse Disease," *Chicago Daily Tribune*, October 25, 1872.

95. "The Horse Disease," *Chicago Daily Tribune*, November 1, 1872. Dr. Rauch had managed the city's past human disease outbreaks and promoted the cleanup of a putrid Chicago River during the late 1860s.

96. "The Horse Disease," *Chicago Daily Tribune*, November 6, 1872.

97. *Report of the Board of Health of the City of Chicago*, 149–50.

98. "The Horse Disease" *Chicago Daily Tribune*, October 25, 1872; "The Horse-Disease," *Chicago Daily Tribune*, November 1, 1872.

99. "The Horse Disease," *Chicago Daily Tribune*, October 31, 1872.

100. "The Horse-Disease," *Chicago Daily Tribune*, November 1, 1872.

101. "The Horse Disease," *Chicago Daily Tribune*, November 5, 1872.

102. "The Horse Disease," *Chicago Daily Tribune*, October 31, 1872.

103. "The Horse Disease: Continued Embargo on Local Transit," *Chicago Inter-Ocean*, November 4, 1872.

104. Henry Bergh founded the American Society for the Prevention of Cruelty to Animals (ASPCA) in 1866 in New York, and Illinois state law defined and articulated punishments for cruelty to animals beginning in 1869. See Janet M. Davis, *The Gospel of Kindness: Animal Welfare and the Making of Modern America* (New York: Oxford University Press, 2016); *Prevention of Cruelty to Animals Act, 1869 Ill. Laws 3*, accessed through Animal Legal & Historical Center, Michigan State University, https://www.animallaw.info/statute/illinois-1869-cruelty-animals-statute.

105. "The Horse Disease," *Chicago Daily Tribune*, November 6, 1872.

106. "The Horses," *Chicago Daily Tribune*, November 8, 1872.

107. "The Horse Disease: Men in Harness," *Chicago Inter-Ocean*, November 5, 1872.

108. "The Horse Disease: The Streets Deserted and Business Brought to a Stand-Still," *Chicago Daily Tribune*, November 3, 1872.

109. "A Text for Sunday," *Chicago Daily Tribune*, November 3, 1872.

110. US Census Bureau, *1900 Census: Volume V. Agriculture, Part 1, Farms, Livestock, and Animal Products, 1902*, 581, https://www.census.gov/library/publications/1902/dec/vol-05-agriculture.html.

111. "The Good Old Times: Streetcar Travel in Chicago Twenty-Five Years Ago," *Chicago Inter-Ocean*, December 21, 1897.

112. "When Horses Were Sick," *Chicago Chronicle*, February 16, 1896.

113. McShane and Tarr, *Horse in the City*, 16, 165. McShane and Tarr argue that the use of horses for urban motive power began its decline around 1900.

114. *Chicago Daily Tribune*, December 31, 1889, 4.

115. "Paris in a Real Panic," *Chicago Daily Tribune*, December 29, 1889.

116. "Brutes Held by Love," *Chicago Daily Tribune*, March 12, 1899.

SUSAN NANCE

three Cattle and Blizzards
Lessons from the Big Die-Up in 1880s Montana

During the catastrophic winter of 1886–87, hundreds of thousands of cattle died of exposure and thirst on Montana's frozen ranges, a disaster that later became known as the Big Die-Up. Before he was widely known as "the cowboy artist" and a much-loved illustrator of scenes from the Old West, Charles M. Russell worked as a range cowboy there. On one job, Russell managed a herd of five thousand head of recently arrived Texas cattle. As Russell told it, when his worried employers sent a letter inquiring about their stock, he was provoked to sketch a starving cow in the snow being stalked by wolves (fig. 3.1).[1] That drawing would be photographed and reproduced many times, displayed in wealthy households and public buildings in Montana. When it first surfaced in the spring of 1887, it was known colloquially as *Waiting for a Chinook*. Russell captioned the sketch *The Last of 5000*.[2]

Many of Russell's depictions of life in the West acknowledged but also romanticized the animal suffering that he and others witnessed. Art buyers and much of the public found that appealing because they imagined the ranching West as a space of resilient men, incredible fortunes (for a few), and inevitable danger.[3] Thus they became nostalgic over images of cowboys riding frantically bucking horses; roping, throwing, and dragging struggling cattle; or engaging in other forms of violence that characterized settler efforts to control or exploit that landscape, the animals, and the people who lived there.

Russell's image certainly expressed empathy for the emaciated, freezing cow at the center of the scene—isolated from her herd, hunching her back in the cold, and lifelessly staring into the empty whiteness of the storm. To many viewers in the late 1880s, the image probably depicted the wolves, or perhaps wild nature itself, as the villain of the scenario. As settlers labored to transform Montana's vast lands into a generous producer of commodities, many took it for granted that winter range "losses" (that is, dead cattle) were unavoidable.

FIG. 3.1. Charles Marion Russell, *Waiting for a Chinook*. Buffalo Bill Center of the West, Cody, Wyoming.

Others would have seen the image as an indictment of unregulated extensive ranching and the speculators who abused their access to land and animals that decade.[4] Both interpretations of the image were possible since Russell did not explicitly ask viewers to think about the people who had put that Texan cow in such a deadly position, nor the attitudes toward the natural world that undergirded settler ambition in Montana.

Traditionally, historians have taken the Big Die-Up as an economic or land mismanagement story, although a "poignant" one, to be sure.[5] On the northern Great Plains, beef was subject to speculation in the 1880s. Well-heeled but naive, absentee investors in the eastern United States and Britain, plus local speculators, competed to mass-produce market steers in places like Montana, while commodity traders in Chicago inflated beef values. Together they created a tragedy of the commons event. Each operation acted in its own self-interest and without coordination, some driving new herds onto drought-stricken, wildfire-scorched, forage-depleted ranges that could not keep so many animals alive. Over the winter of 1886–87, thousands of already run-down cattle died in historic, cruelly frigid storms that paralyzed much of the Great Plains.[6] In spring, as cattle growers shipped the survivors to urban slaughterhouses, the run-down state of those animals caused beef prices to collapse as opportunistic packing trusts offered little to cattle sellers made "desperate" by the crisis.[7]

Recent scholarship has revised this story to deemphasize simple, uniform overstocking and range depletion as the cause of the disaster, to argue instead for the combined role of investors' and ranchers' relish for risk and their profound cluelessness about the variability of the West's landscape and weather.[8] Nonetheless, all these explanations have so far failed to reckon with the historical significance of the animal welfare costs of the disaster.

To tell a concise story that reveals broader historical patterns in how animals have confronted anthropogenic environmental change as well as human population growth, mobility, technology, and ambition, this chapter zooms in on Montana in the years just after the cattle trade had finally reached that last corner of the northwestern Great Plains. It explains what the Big Die-Up was like for cattle on the ground as they struggled, often unsuccessfully, to find water, forage, and shelter in frigid temperatures. There, investors, railways, and packing trusts externalized the costs of beef production onto ranching operations, certainly, but also onto cattle themselves. They left cattle to roam unfamiliar ranges with little to no support, such as hay supplies, accessible and dependable sources of water, and shelter from storms. Corners cut on the range produced unimaginable suffering and death for creatures who circulated in the food system as marginalized actors. Subsidizing the beef business through their ability to tolerate deprived conditions at times, reproduce and survive when they could, thousands dying when they could not, the only currency that mattered to cattle was their own well-being. Indeed, out on the land, cowboys and other rural Montanans witnessed that reality firsthand in the disaster and its aftermath, later recounting its unsettling if not traumatic nature.

This Big Die-Up story is also about the failure of attempts to salvage the situation. During blizzards, cowboys worked in dangerous, often miserable conditions to help cattle survive. In many places they were unsuccessful. Humane advocates labored fruitlessly in those years to change the industry by attempting to enlist consumers in public demands and legislation to protect cattle from abuse and neglect. In both cases, people noted the relationality among vulnerable people, fragile lands, and exploited cattle, whether cowboys putting themselves at risk to protect cattle in an inhospitable valley or section of plain, or consumers trusting their health to disinterested actors controlling the now continent-wide food system. Thus, this chapter explains livestock-growing in Montana, the nature of cattle behavior on ranges leading up to the 1880s, how cattle experienced the winter of 1886–87, and to conclude, why in the aftermath of the disaster the few incremental changes that did come were driven only by industry, not consumers.

Charlie Russell's likeness of that starving cow, whoever she actually was, is a reminder that, although some in Montana knew cattle and their suffering directly, the majority—including the investors who wrote to Russell about their stock—were entirely alienated from that reality. Repeated animal crises have taken this form. Investors and consumers had only secondhand knowledge of remote locations of production while pro-market colonization and technological development encouraged many people to internalize the myth of human ecological autonomy. Long before people thought to speak of "anthropogenic" changes in the United States, Montana's cattle inhabited a spot on a timeline stretching back several centuries. Modern cattle ranching was rooted in the global, commoditized circulation of bodies, discussed by Josh Kercsmar as an aspect of the imperial context into which the United States was born. Like colonial sugar planters, cattle ranchers externalized the costs of production onto living beings, drawing some criticism from reform groups but no real solution. Most took for granted or barely noticed how animals subsidized Americans' activities with their ability to survive for periods of time in deprived or dangerous conditions—until they could no longer. A decade later in Wyoming, elk would find themselves in the same situation, but not because they were incapable—they resided in their home territory, after all. Confronting colonization on the old bison lands to the north, elk and other wild animals were suddenly hemmed in—like cattle caught on a barbed-wire fence in a snowstorm—by an influx of settlers. Elk lost much of their autonomy in the process, again inspiring a minority to advocate for them and look for ways to help them survive. That impulse in turn imposed labor onto local settlers and their agricultural systems, which were simply not up to the task, nor designed for such responsibilities.

To see this disaster from the ground up, we should begin from the premise that the Big Die-Up was not an unpredictable accident or anomaly. Instead, it was the high point of a cascade of systemic problems that those terrible winters revealed, or should have revealed, to everyone. The problems emanated from the combined impacts of colonization, long-distance capital investment, and the growing power of railways and meat packers to define how animal agriculture functioned.[9] Although the industry might be highly efficient at the slaughterhouse, on the ranges cattle ranchers and investors understood individual cattle to be expendable and so tolerated many practices that proved wasteful from a business point of view, if not a humane one.

Although people had been raising beef cattle in North America since Spanish ships had carried the first Iberian cattle to the continent centuries earlier, not until after the US Civil War would a truly continental beef market appear.[10] Beginning in the 1860s in Texas and the Southwest, entrepreneurial men discovered that money could be made by buying animals, moving them to "free grass" (that is, grasslands recently confiscated from Indigenous tribes), allowing those cattle to graze and reproduce, and later shipping them for slaughter to a growing population of beef-eating consumers across the continent. As accessible and more productive land was taken up nearer those consumers, cattlemen looked farther and farther afield north and west to find more land. Finally, in 1881 the Northern Pacific Railroad arrived in Montana, and Miles City became the state's main livestock railhead town. Montana, although remote, held some of the last available rich grazing land in the United States.

The animals drawn into the new cattle trade in Montana were long-legged, long-horned Spanish-descent criollo cattle born wild in Texas or northern Mexico, including the famous Texas longhorns.[11] With their feral culture and individual curiosity, these cattle passed survival skills from generation to generation through experience at large on the arid lands of the Southwest. According to the travelers and soldiers who occasionally sought to kill and butcher them in Texas, in order to avoid contact with humans, criollos hid in dense thickets and brush, forging tunnel-like openings that a man on horseback could not navigate. Cows and calves were especially wary, and Texans reported seldom seeing them up close.[12] When people still found ways to confront them, they revealed themselves to be "wilder and more apt to attack" than the bison or other ungulates.[13] It was a truism that one must remain on horseback to venture safely across the land or risk unexpectedly encountering some of these creatures who might not charge a person on horseback but were curious and nervous around a human on foot.[14] In Texas and the Southwest, cowboys admired criollo cattle for their self-sufficiency, their hardiness, and the challenges they presented. Such qualities only embellished the skills of the *vaquero* who knew how to work with them.

In the scramble to bring cattle into northern ranges, people also imported northern European shorthorn cattle descended from heavy beef breeds and raised in intensive settings in the Midwest, where people kept cattle on small ranges and supported them with hay.[15] Some of these experiments sought to mix longhorns with heavy beef breeds from northern Europe to make them easier to drive and handle by diluting their "wild" and autonomous ways.[16] Cattle autonomy and self-sufficiency was crucial to hands-off ranching, premised as it was on minimal supervision of cattle or inputs of labor on the range to keep them fed, watered, and safe from the weather. It also prevented cowboys from supporting cattle during weather crises since, wary of human management, such cattle often misunderstood or ignored attempts to move them to safer ground.

Driven on foot for weeks or months, most of these creatures found themselves reassembled in unfamiliar herds where they created temporary communities according to their own instincts and experience. These efforts focused on preserving group cohesion and social hierarchies, forging specific relationships between individual cattle, and protecting calves. Cattle lived in cow-calf herds with dominant bulls and steers in "bachelor" groups slightly on the fringes. Dominant bulls ventured among the cows only when they smelled some in estrous, or perhaps to display status by tossing some dirt on one's back with a hoof and snorting an offer of a head-butting challenge to other bulls lurking about. When being trailed or collected, cowboy herding compressed that social organization, which probably caused some of the conflict cowboys reported among herds, since there was a greater likelihood that bulls would irritate or challenge one another in close quarters.[17]

Once stationed on a home range in Montana, cattle organized themselves in "bunches," as the cowboys called them, of about twenty animals.[18] Even in the best of times, it was hard work traveling to find forage or water in those groups. In droughts or storms it could also mean that cattle "walked off their flesh" by traveling long distances to find water or an ungrazed area.[19] Although the animals saw cowboys and horses at times, ranch operation managers and cowboys left them to fend for themselves for long periods. Only during spring roundups for branding and fall roundups for capture to go to slaughter did anyone rope or handle them. One cowboy recalled that such rough handling—as necessary as it was to the production of market steers—often produced "a loss in flesh [that] . . . cut into profits."[20] Many of the techniques cowboys used to move or control cattle were frightening or painful—yelling, firing pistols, or running at them with horses to set off flight response.[21] Extensive ranching was built on inexpensive, hands-off management paired with painful or frightening handling at roundups that taught cattle to be wary of human contact, or "wild," as the cowboys would explain it.

The men who did this range work were a diverse, working-class community of settler, Indigenous, Black, and Hispanic men who mostly learned on the job how to work with cattle. As a rule, there were not enough men on the crew. Employers endeavored to keep costs down, routinely paying cowboys only thirty dollars per month. They faced rural labor shortages in the 1880s and 1890s, but did not raise wages to make the work more appealing.[22] The precariousness of range work was emblematic of a cash-starved and unsafe work environment that highlighted the relationality of animal and worker exploitation. Laboring in remote locations with no medical support, insurance, means of communication, or—less anachronistically, in an age of considerable union activity—agricultural labor unions, cowboys were vulnerable to employers and the environment in their own right.[23]

Documenting that work, the few published memoirs of cowboy life were generally authored by settler men who celebrated their time in the saddle and the opportunities ranching provided. While they give us a glimpse of how cowboys observed cattle and what they understood of cattle communication or behavior, they cannot be taken as a comprehensive account of all cowboy attitudes or knowledge. It is likely, as Josh Specht proposes, that most rural range workers, especially people of color, "may not have shared that optimism" about the industry that famous cowboy memoirists voiced.[24] Still, although some cowboys found cattle to be stupid or exasperating, these men spent many hours observing cattle behavior, and in more than a few cases they did so with empathy. Many seem to have proceeded from the basic assumption that cattle were intentional, emotional beings with the ability to learn from experience. Collectively, their words reflect a skepticism about the industry when blizzards arrived.

Range workers daily witnessed cattle endeavoring to build interreliant communities while traveling to find forage and water. "I firmly believed that cattle could talk to one another," reported Hubert Collins in his range memoir in discussing the various "baw," "roar," and "ow-o-o-o-ou" calls he heard and saw cattle direct to one another. "During the long days on herd I would sit on my pony with leg thrown over saddle horn, asleep some of the time, and rouse at any unusual disturbance, including the bull fights which occurred several times daily." Collins was multilingual—in human languages—but also claimed over time to see that cattle used "inflection, tone, manner of delivery [to] give very different meaning to the same general sound."[25] In general, the cattle he tended disliked being left alone or separated from their bunch if driven into a larger herd. When driving steers to the railhead, frequently the cowboys would see a lone, "dragtail" steer who had lost track of his "mates." "Wild cattle love company and if their folks are driven away, leaving one to solitude, that one will do all he can to join the crowd, and he will talk about it," Collins explained. "Before he has discovered the missing ones, he trots along, never pausing except to voice his inquiry of 'Where are you?' Once he discovers the missing ones he proceeds at his fastest trot, scolding the while that he should have been left behind. When he has found his friends and relatives again, he falls into line, giving a final satisfied bawl."[26]

In particular, Collins and other cowboys commented on the communication between individual cow-calf pairs, who recognized one another even in large herds. Collins, perhaps with a bit of romantic relish, described the cow's devotion to her calf as just one of the "human traits" cattle displayed.[27] Among cattle, social organization and the ability to find and identify individuals in a herd is determined by smell, not vision, touch, or hearing. As such, many cowboys may not have realized that their cattle employed scent to keep track of one another and evaluate the relative contentedness or stress in the herd.[28] In spring, calves could be seen "dashing here and there in playful jumps," thus becoming lost in the herd and calling for help, Collins remembered. "Then mother starts to the rescue, giving a warning to all other elders, which says, 'Clear out of my way, for I am *going* to my child.' Then she plunges straight toward the child's voice and they all make way for her. I have heard them as plainly as if they talked English."[29]

The mysterious link between cow and calf was perhaps most obvious in the cow practice of "caching" her calf. After giving birth many miles from water, cows would hide their young calves under some brush, making the trek to water, drinking, then returning to nurse, thereafter repeating this practice until the calf was strong enough to travel. Many cowboys reported with astonishment seeing young calves lying alone and motionless in some remote spot. No amount of prodding or urging would get them to move or even change position. In time, a cow would appear in the distance, trotting as though on a mission, "nose to the ground."[30] Hearing the cow call with a particular "baw," the calf would suddenly spring to life once she drew near, then attach to an udder.[31] Even in mild weather, these social connections and self-sustaining behaviors would be tested by the vastness of the enormous ranges believed necessary to support herds in operations using the cheap, hands-off mode of management. Or, rather, cattle's ability to stay healthy was tested by the risky choices people made to put cattle on particular sections of Montana's landscape although they had limited knowledge of its particularities and what the weather would bring. While much was made of the risk taken by local and international investors, it was cattle and, to a lesser degree, range workers who had the most to lose.

Long-term residents of Montana well understood the danger of the state's variable landscape and unpredictable weather patterns. The state consisted of two main ecosystems: the forested, mountainous strip on the west side and a vast expanse of treeless plains to the east. Such an expanse of land made accurate generalizations about weather or land condition impossible.[32] Plenty of immigrant ranchers in the American West had ignored such wisdom when they encountered it in the late 1880s. Drawing upon their experiences of cattle growing in the East or Britain, in the fall of 1886, many ranchers supplied insufficient shelter and stockpiles of hay, then watched helplessly as their cattle died the following January and February.[33] Tribes of the high plains took seriously Montana's periodic droughts, rainstorms with flash flooding, and unpredictable, deadly cold snowstorms. Although they had the knowledge and in their recent past had practiced it, many Indigenous people in the region had given up on formal settled agriculture in favor of a nomadic life that allowed them to stay mobile in the face of fluctuations in weather, snow, or rainfall. Previous generations had followed the bison herds as they traveled to the best grasses, occasionally evacuating exposed prairies or heading south when frigid temperatures were imminent.[34] Prospering consistently on the land was a product of experience and adaptability for human and nonhuman alike.

On a small scale, members of local Indigenous tribes knew it was certainly possible to raise livestock in Montana. They had been raising horses there for over a century. After many generations in the state, their horses knew to eat snow for water and to scrape through snow with a hoof to find frozen grasses to eat. During storms, they traveled intentionally in small bands to find shelter in canyons, among trees, or on the leeward side of hills, where winds had blown the dry prairie snow away to reveal edible frozen forage.[35]

Likewise, small numbers of beef and dairy cattle had been afoot on the northern Great Plains and Montana territory since about the 1840s. Initially, some appeared just to the south in what became Wyoming and Idaho, driven by overland migrants traveling with oxen or perhaps a shorthorn beef breed or a dairy cow as they camped their way toward Oregon. Some cattle abandoned their owners or were abandoned by them, turning feral or being captured by later miners, missionaries, or migrant traders, some of whom became the earliest Montana cattlemen. Later, in the 1860s, as local Indigenous tribes, settlers, and the US military came into armed conflict, the federal government sent teams with small herds designed to supply forts and other government outposts. Next came homesteaders who likewise raised small bunches of northern European shorthorns at their properties. With the founding of Indian reservations in 1880s Montana, the US Bureau of Indian Affairs put herds on reservation lands in (often vain) attempts to prevent men from leaving to hunt cattle and game, not including bison, who were functionally extinct by then.[36]

Before the beef bubble of the 1880s, people raised their cattle close at hand, primarily for local or perhaps regional consumption (for instance, driving them from western Montana to eastern Oregon), since transporting live cattle to slaughter was costly. In regions where geography allowed, people employed seasonal grazing, sending cattle to more exposed prairie ranges in the warmer seasons, then herding them in nearby hills or forests that could provide some shelter over the winter. Rather than a high-volume approach, which gambled on minimal winter losses and shrugged at a certain level of cattle misery and death as an unavoidable cost of business, smaller-scale ranchers more often had the intention that each animal would survive until spring in the best possible condition.[37]

At this point, cattle were an invasive species. Along with equines introduced to the West by Indigenous horsemen a century and a half earlier, concentrations of cattle in any location put pressure on old bison lands suddenly asked to support denser concentrations of animals. Here cattle were part of a consciously colonial environmental transformation. Many settlers and agents of the US government encouraged the transition from bison to cattle as a way to disrupt Indigenous relations with Montana's landscape and, not incidentally, the economic autonomy and food security that came bundled with Indigenous ideas of animal personhood. Broadly speaking, those philosophies espoused an idea of mutual responsibility that acknowledged equal vulnerability in the environment among people, land, and animals, all of which was steeped in realism earned by many generations' experience living in the West.[38] Indigenous understandings and experiences of raising livestock conflicted with the hubris and tolerance for risk that many settlers and opportunists in the region had internalized. There was a political and ecological shift underway, with cattle serving as pawns in nation-building.

By the early 1880s, the situation for cattle on the Great Plains and in Montana had begun intensifying. At that historical moment, those Montana grasslands were an ecosystem undergoing anthropogenic transformation away from the "bison ecology" network of plants, animals, and insects that had been in place for a number of centuries. Things looked bleak by the early fall of 1886. Yet, even with limited capital and struggling stock, many cattlemen nonetheless looked at ballooning beef prices and wagered that, if they could hold fast, surely they would have herds of great value by spring. Such bonanza talk promoted a scheme by which the cattle did most of the work and risked their lives to serve a nascent consumer society built on domestication of the environment. Settlers had for several centuries already enacted conscious destruction of ecological diversity in favor of preferred species: cotton, rice, and tobacco monocultures in the South, cattle kingdoms on the Great Plains. These agricultural empires predated the pork monocultures that Mary Trachsel explains later in this volume, wherein an alliance of industry, government, and consumers seeking inexpensive products together transformed Iowa's ecology. Montana ranchers could only dream of such transformations.

One legend explained that the beef bonanza began just after some freight drivers abandoned two unworkable oxen teams in heavy snow in northern Nebraska. The men returned the next spring, and "instead of bleached skeletons they found the oxen themselves, sleek and contented," as one cowboy told the tale.[39] The mistake people made thereafter was to imagine that exception as normative and—on the basis of minimal experience—go all in, speculating in imported herds and searching for new land to exploit. A cottage industry in how-to-get-rich-on-the-plains books and newspaper stories also appeared, enticing entrepreneurial men to import cattle into the northern Great Plains.[40] Some argued that colder weather caused cattle to put on more weight, such that they would earn more at market by grazing in Montana or Wyoming than in Texas or Colorado. Thus, those northern territories were the "choice ranges for breeding and fattening cattle."[41]

Beginning just north of Kansas, placement of herds had spread, year by year, north and northwest, finally arriving in Montana in the mid-1880s as people saw that the low-hanging fruit of free, nearby ranges (grabbed by earlier ranchers or settlers) was depleted. That is, when the more convenient land was taken, cattlemen and their hired hands drove cattle north and west, resorting to more remote and less familiar, more risky settings. "Always and eternally, the weather," wrote state historian Joseph Kinsey Howard of how Montana's climate and landscape so often thwarted settler ambition—although to no avail. "Always in Montana the bitter conviction [is] that its vagaries are exceptional and malevolent, though they were so unexceptional and periodic," he explained of the settler habit of interpreting harsh winters and years of scant rainfall as anomalous and surely not to be repeated.[42]

Many of the men who brought in cattle to graze were not completely credulous, of course. Still they sought to create the best-case scenario in an increasingly precarious and crowded public lands context. In finding a place to overwinter a given herd, the cattlemen, their agents, and hired men understood that cattle needed shelter in winter to stay alive but also to minimize weight loss. Still, competing over land with homesteaders and one another, there were compromises to be made. Ranchers often chose land without shelter but supplying some water and "well grassed," since cattle would certainly die without those, then hoped the winter would be mild that year. "Ground interlaced by hills and hollows offered to the animals in winter not only patches of grass devoid of snow, but also screens from bitter winds," advised cowboy Philip Rollins. "Dangerous as were winter's storms in open country, they were not as perilous as were summer's droughts on arid ranges," he argued in weighing the risks.[43]

As winter descended in 1886, in many cases, men simply miscalculated and discovered after a period of months that forage was running out sooner than expected. With no fresh land available nearby, there was nothing that could be done to save a herd. Montana's newspapers reported on the fate of individual herds since two herds might experience very different outcomes although stationed near one another. Or newspaper reports of fat and fecund herds planted in the papers might simply be fraudulent.[44] In an age before ecological surveys or even the most rudimentary range science studies—which have, in any event, routinely fallen short of accounting for the limitless combinations of various plants, water, soil, weather, grazing animals, and wild animal and human activity that determine the carrying capacity of a given acre—ranchers employed a combination of past experience, guesswork, and hope to evaluate ranges.[45]

The wild and self-sufficient reputation of the new cattle plus the historical success of bison there possibly colored settlers' and investors' expectations that cattle would prosper on most ranges. Previously, bison herds had not overgrazed western grasslands nor perished en masse because they encountered no fences (which began appearing in Montana in the 1880s and restricted cattle to areas of drought-stricken or grazed-out land) and had the stamina and instinct to travel longer distances to graze. Thus, bison spent less time than cattle standing in fragile riparian zones, so they distributed more widely hoof damage that suppressed grasses around crucial bodies of water. These were some among many behaviors, which supported the larger "bison ecology" of grasses and animals linked to that keystone species, that beef cattle would fail to replicate.[46]

Montana's blizzards had always been notorious and unpredictable. Then, beginning in 1881, winter crises similar to the one causing the Big Die-Up had occurred repeatedly on the Great Plains.[47] "At least 1,000,000 female cattle have left Texas for the Northern grazing ground," journalist and explorer Frank Wilkeson had warned in a *Harper's New Monthly Magazine* piece written in the spring of 1886, just months before the Big Die-Up storms began. Ranchers and industry observers estimated that there were approximately 900,000 cattle in Montana, double what had been there just a few years earlier.[48] The work of counting or even estimating cattle numbers on remote ranges was rife with conjecture and frequent misrepresentation since there was no way to monitor animals over Montana's vast landscapes while also accounting for births and deaths.[49] Importantly, the inability of ranch managers to make accurate herd counts was a warning ignored that cattle were not being sufficiently supported to survive. "Where are they?" Wilkeson asked rhetorically. "The bones of thousands of them lie bleaching on the wind-swept flanks of the foothills of mountain ranges; . . . they lie in disjointed, wolf-gnawed fragments on the arid, bunch-grassed ranges . . . low-lying monuments of man's inhumanity to the dumb animal he has arrogantly assumed charge of." Wilkeson was right. Although advocates for the radical ecological transformation of the state spoke and wrote about the march of colonization, capital, and speculation as inevitable, progressive, and utopian, there were always those who recognized folly when they saw it. Said Wilkeson, "The country these cattle are held in has been in the white man's possession for but a few years. The [settler] men who have accurate knowledge of its climate can be counted on the fingers of one hand."[50]

As cattlemen and property lines limited where and how cattle could move, in many places, local ecosystems labored under concentrations of more stationary animals. In the fall of 1886, many said that ranges near the base of the Rocky Mountains were overgrazed before any snowfall that year because too many homesteaders had forced herds off the plains into the hills and mountains.[51] In other places, wildfires or grazing herds of sheep reduced forage in the fall when there would be no time for it to sufficiently regenerate. Other regions had become riddled with toxic or inedible plants that sprouted where cattle grazed the old bison grasses past the point of regeneration.[52] The *Billings Gazette* reported on the dark humor that had begun circulating about the growing crisis on certain ranges, saying about some Judith Basin land in central Montana, where well-known rancher Granville Stuart had a herd situated, "The Maginnis range was aptly described . . . recently as being 'a very good one, except that there is neither grass nor water to be had on it.'" Stuart admitted that the land was so run down that to leave his herd there "would simply mean their extermination."[53] Indian reservations were only barely established in the early to mid-1880s and remained unfenced, with their boundaries only symbolic to resentful or desperate cattle growers.[54] "This is rough on the cattle men whose half-starved animals have been tempted by the sight and smell of the luxuriant herbage on the reservation, to stray across the boundary line and live in clover while they may," said an editor at the *Billings Gazette*. A noted supporter of permitted grazing of settler cattle on reservation land, in December 1886 as the storms began, the paper regularly noted periodic fines levied by Bureau of Indian Affairs officials against cattlemen who "allowed their cattle to run" unauthorized on such land.[55]

Winter storms would again temporarily interrupt cattle behaviors designed to promote herd social dynamics, cow-calf relationships, watering, and the keeping of meat on one's bones. Storms occurred around the state in late fall and continued until just before spring, punctuated by a crippling blizzard on Christmas day that "never really let up for sixty days," cowboy Edward Abbott remembered. Then warm chinook winds came from Alberta, melting the two feet of snow on its surface just enough to form a thick crust of ice once temperatures dipped well below freezing again. On February 3–4, 1887, another terrible blizzard set in, and "it was hell without the heat," Abbott said.[56] Snow events of that magnitude and timing could negate many months of work by range cowboys in keeping a herd from being killed by wolves or human hunters and surviving whatever dry or cold weather preceded it.[57]

For cattle, life on a winter range in the midst of rapid ecological change—of which they were just one element—driven by human ambition and speculative investment was routinely miserable and often deadly. Over several previous years in Montana, imported cattle born in Texas, Mexico, or intensive contexts in the Midwest first revealed they were struggling by way of lower birthrates and a high incidence of aborted calves.[58] Even the boosters admitted that "the losses on winter-dropped calves, and on young yearling mothers, constitute in Northern States the principal percentage of loss," as Baron Walter von Richthofen, a widely read beef bonanza author, explained to would-be investors, for instance. Yet, if one invested men and hay to watch and feed those cattle, one could keep those "losses" to about 1 percent, or up to 3 percent in a worst case, he promised.[59] Cattlemen engaged in just such calculations. Knowing that not every animal would survive, that some would starve or freeze to death, cattlemen estimated how little an input of labor and forage support they could supply without going broke from expenses and without seeing herds collapse if the winter was long or especially brutal. Many certainly gambled that, while the fellow or operation down the road might be headed for tragedy, they would somehow outsmart nature.

Winters taxed cattle because they had no time to adapt their behavior—if that was even possible—to the particular combination of sparse forage, long distances, limited water, and restrictive fencing on any particular range. J. Frank Dobie reported of the wisdom in northwest Texas that longhorn cattle driven north early in the spring who survived an initial winter would prove indestructible thereafter. It was "pioneers" (criollos driven north late in the year) and "barnyard stock" (northern European heavy beef breeds raised in captivity in the Midwest) who routinely perished up north.[60] Said cattleman Granville Stuart of the contrast in Montana in 1886 and 1887, "The herds that were driven up from the south and placed on the range late in the summer, perished outright." He speculated that, even among the more seasoned cattle in Montana, in many places a death toll of about half or three-quarters was common that year.[61]

Once at work on the land, cowboys discovered that the new cattle, so self-sufficient in many respects, displayed behaviors that foiled attempts to support their overwintering. They told stories of being frustrated in their attempts to help, giving up, and heading to a remote range cabin to wait the storm out.[62] Once temperatures dipped, these cattle consumed snow as a source of water only incidentally, while eating frozen grass that was covered in *just enough* snow—not blown free of snow by Montana's constant winds nor buried under more than a few inches. When cowboys artificially created waterholes in frozen lakes and ponds or creeks, chipping out openings and herding cattle to them, the cattle gratefully drank the water but refused to move thereafter. Soon consuming all the grass nearby, "they will huddle up and silently starve, hanging about the watering place and refusing to go out onto the hills and feed," said Emerson Hough.

In heavy snowfall, a herd might also be hemmed in by deep snow and unable to get out to higher ground. "Sometimes trails are broken up from the valleys by the cowboys, who ride their plunging ponies back and forth and finally form a roadway over which the cattle can be driven out from the death trap into which they have come and which they refuse to abandon," Hough explained of the exhausting, even desperate work of trying to save a herd.[63] "Think of riding all day in a blinding snowstorm, the temperature fifty or sixty below zero, and no dinner. You'd get one bunch of cattle up the hill and another one would be coming down behind you," explained Edward Abbott. "It was all so slow, plunging after them in the deep snow that way; you'd have to fight every step in the road."[64] Sometimes, once all the grass was gone near a water source, hungry cattle resorted to eating willow branches and other indigestible plant matter. "This caked their stomachs with dry murrain which couldn't be cast off, and many died a horrible death," remembered cowboy Ramon Adams.[65]

Among the Ampskapi Piikani (Piegan Blackfeet) and other local Indigenous people, 1885–86 would become known as "many cattle died winter."[66] This was a plain and melancholy formulation, acknowledging the essential tragedy of what beef investors and their managers allowed to happen. Unlike the "Big Die-Up" euphemism, here there was no flippant or darkly humorous likening of what cattle had experienced to traditional "round-ups," namely the (often romanticized) cowboy work of catching and branding cattle in spring and the lively fall community ritual of cattle auctions and socializing.

Notameohmésêhese (Northern Cheyenne) historian John Stands in Timber was born in 1882, just before these events and two years before the founding of the Northern Cheyenne Indian Reservation in southeastern Montana.[67] Years later his uncle, John Crazy Mule, told him about the winter of 1885–86. The snow had been incredibly deep that season and some among the tribe were still living in teepees. Making their way to the Indian agency depot at Lame Deer to retrieve routine food supplies (a trip made necessary since Notameohmésêhese were now banned from traveling to hunt as they had in the past), John Crazy Mule recalled how a crew of men "had to shovel most of the way to get through." It took two days each way. "They said the bones of the cattle that starved that winter could be seen years afterward in a place along the river between Birney and Ashland. Two or three hundred had crowded in under some sand rock cliffs and died."[68]

The cattle described by John Crazy Mule were typical of those recently arrived in Montana: struggling unsuccessfully with sudden storms and impossibly cold weather using the habits and instincts they had used to find water or shelter on a different landscape. Traveling together, in blizzards they would abandon their familiar home range (a valley, plain, or set of hills in which they were stationed and held within a larger range) and "drift," as the cowboys called it, with their backs to the wind.[69] "Cattle would drift day and night in a blizzard till it was over. You couldn't stop 'em; you had to go with 'em or wait till the storm was over, and follow," Ramon Adams recalled. The habit was so predictable that some referred to it as "winter drift," explaining it as an attempt to escape severely cold winds and frozen water or forage, although it could also occur in summer droughts that similarly limited forage or water. "With the stingin' sleet cuttin' into their eyes and faces, it was a waste of time to try and stop 'em," Adams continued.

> Shortly they became hungry and numb, thin and gaunt. Glistenin' icicles hung from their mouths, ears, and eyes. On and on they plodded before the

storm, lowin' their misery to a seemingly empty world, heads lowered, hairy backs snow-covered and crusty, their eyes but dark holes peerin' out of an ice pack. . . . With heads down and tails tucked, they stumbled on. With legs and hocks sore, hairless and bleedin' from walkin' in crusted snow, they left a trail of blood and frozen bodies of them that got down and couldn't get up. Here and there was a thin-flanked cow with a shiverin' rough-coated calf trailin' at her heels, maybe wonderin' at what cruel world he'd come into. Also there was an occasional humpbacked yearlin' with little nubs of horns tellin' that they were jes' comin' out of calfhood.[70]

In Adams's account there is a sense that many who witnessed such pathetic scenes felt sadly helpless, if not traumatized by how vulnerable and miserable were these creatures whom men had put in harm's way.

When a storm set in, many operations sent out crews of riders to try to salvage the situation, but often it was not enough.[71] Cowboys sometimes had to abandon such herds for their own safety, leaving the cattle behind knowing many would die.[72] In other places, owners had been pushed out of prairie areas and valley bottoms by homesteaders, so they herded cattle up into the hillsides and subalpine mountain areas, from which cattle drifted downhill looking for water. At one point in February 1887, after many days in unforgiving conditions, cowboy John Barrows thought to himself that "the storm had continued beyond reasonable bounds." With the blizzard "still raging," he recalled: "An exodus of cattle from the upper valleys set in and for two or three days there filed past our place along the river bottom a discouraged and hopeless procession of cattle, hungry and freezing, driven by their distress. It was bitterly cold, with a strong gale blowing from the north, truly perishing weather. I could compare the flight to nothing but pictures I had seen of the retreat from Moscow."[73] Here Barrows refers to the October 1812 retreat of Napoleon's crippled forces from a disastrous invasion of the Russian city, which was depicted in a number of famous paintings showing exhausted, freezing men and emaciated horses stumbling and collapsing in the snow.[74]

Initially, settler accounts of the disaster seemed to focus on the traumatic nature of the event to observers who encountered surreal and saddening scenes, often encountered with little warning. Around the first of February that winter, Joe Ziminski was traveling just northeast of Billings. An immigrant from Poland, he made a business of supplying firewood by venturing out to the countryside to find snags, bucking and chopping them, then delivering them to town for sale.[75] With the temperature well below freezing—in the twenties Fahrenheit, the papers said—he arrived at Alkali Creek (today the site of a bicycle and hiking trail with attractive views of the Billings skyline) to a bizarrely gruesome sight. "There are many dead cattle up Alkali creek, and in that vicinity," he told staff at the local paper. "Some of them are standing up, and their horns project over the snow drifts where they gave up the struggle for existence."[76] Most histories of this event recount similar firsthand reports; there seems to be an endless number of them in letters, memoirs, and newspapers of the period, describing cattle found dead where they had huddled in waterways or became stuck piled up against barbed-wire fences and other barriers. Cowboy Emerson Hough noticed that one would find in a mass of dead animals a cow and calf together, then nearby all the yearlings, and farther afield bulls.[77] That is, bunches of cattle died in spatial patterns that sadly displayed to observers the social relationships they had cultivated while alive.

These deadly storms were especially hard on Texas cattle brought to the state only the previous fall, so not considered "native" yet. "Losses from drowning and exhaustion in drifts will equal those from cold," said an editorial in a Benton, Montana, newspaper.[78] Indeed, locals observed that these cattle were less likely to eat sagebrush, less hearty, and prone to losing weight or dying suddenly.[79] "Those that were left alive were poor and ragged in appearance, weak and easily mired in the mudholes," wrote Granville Stuart in an oft-quoted passage. "A business that had been fascinating to me before, suddenly became distasteful. I wanted no more of it. I never wanted to own again an animal that I could not feed and shelter."[80] To many in Montana, the scenes in the spring of 1887 were reminiscent of the rampant killing of the bison fifteen years earlier when market hunters might shoot many dozens in a day to feed the robe and hide markets, or just for sport.

Some acknowledged the horrors of the disaster but were not deterred. One Scottish immigrant turned cowboy, John Clay, reasoned, "The cattle cyclone like an Alpine avalanche was no respecter of persons. It hit the just and the unjust. It was the protest of nature against the love of gain, against greed, mismanagement and that happy-go-lucky sentiment which permeates frontier life." Unlike Granville Stuart, Clay remained optimistic. "And yet," he continued, "what would the West have been without the trapper, the miner and the cowpuncher, the pioneers in that wondrous country which pours wealth through its products into an already overflowing reservoir?"[81] Expressing that colonially confident ideology that seemed to drive so many in the West to persevere with problematic attitudes and risky practices, to him the welfare and experience of cattle provided not a warning to change course but simply a colorful episode in the settling of the West.

The Big Die-Up was pivotal in the environmental and economic history of Montana and, to a lesser degree, the history of beef cattle welfare. The Die-Up convinced those who stayed in the business that it must shift "from an adventure into a business." To some degree, investors and ranch managers would thereafter seek to externalize fewer of the costs of the business onto cattle by way of reduced welfare and excruciating, slow deaths by thirst, starvation, or exposure. Increasing support to cattle required ranchers and their cowboy employees to invest in water supply systems, fencing to keep cattle within reach of supplies, and stockpiles of hay to weather difficult winters. They also invested in expensive beef breeds like Herefords. No longer relying solely on wild "scrub stock and longhorn" cattle who produced limited amounts of tough, stringy meat, investors and managers were perhaps less likely to treat their stock as a high-volume, low-profit-per-unit commodity.[82] Still, like Chicago's workhorses a decade earlier, mine mules a decade later, and Iowa's captive hogs a century later, range cattle nonetheless inhabited a world premised on the disposability of animal life.

Consider again the empathy symbolized by Charles Russell's famous drawing *Waiting for a Chinook*. For two decades by then, humane organizations had been pointing out that extensive ranching was a form of profit-driven neglect in the northwestern Great Plains. Modest numbers of voices in Montana and across the country decried the neglect and suffering experienced by range cattle. Their writing appeared in animal advocacy periodicals, western newspapers, and high-profile publications like *Harper's New Monthly Magazine*. Henry Bergh, founder of the Society for the Prevention of Cruelty to Animals, had warned cattlemen that "the laws of nature cannot be violated with impunity," as he labored to give animal suffering some kind of value in a market system where every other aspect of an animal was monetized. Yet he and other animal advocates failed to convince the bulk of the meat-eating public that cattle suffering or neglect had any real financial or moral value. Nor did they effect the passage of any legislation ensuring basic support to range cattle that would be enforced.[83] Cattle on remote ranges out west were simply too far from consumers' daily experience to elicit the same attention as other animal cruelty crises, such as the equine epizootic fifteen years earlier.[84]

For the absentee cattlemen and many small-scale local cattle growers—some of them speculators in their own right—that deadly winter was reason to quit the business.[85] For other locals, it was simply a case of bad luck. Many ranchers were "next year people," as Walter Prescott Webb's classic account of this history explains. Their optimism was essentially ideological, employed to reconcile contradictions between what people believed the West could provide and actual experience. Aspiring ranchers and farmers who appeared thereafter on the horizon and settled in (however precariously) during better seasons continued writing home that "the country is 'getting more seasonable.'" Webb explained, "Always that fiction, the expression of a vain hope, asserted itself in the fat years of the West."[86] Although many people understood the suffering of Montana's cattle, and even felt bad about it, they did not weigh the experiences of cattle against the promise of economic development or individual fortune.[87] Instead, even after the Big Die-Up, people doubled down and tried again.

Since cattle experience had no economic value to consumers and only limited importance to producers, the national beef infrastructure would continue to develop with only incremental changes. Those industry improvements helped people ignore cattle experience, believing that winter "losses" and the cattle misery that came with it were unavoidable in the work of asking arid, remote, often unknowable, occasionally deadly frigid lands to feed a growing human population. Shortcuts in the system were externalized to the land (through destruction of waterways and crucial native grasses) and, in time, the government (through federal management of public lands and grazing leases), certainly. The costs of beef production were also externalized to cattle themselves—the most vulnerable players in the system. Cattle growers and beef consumers did not pay the full price of meat production, but asked cattle to subsidize the industry by struggling along in a place and time for which they were not designed.

NOTES

1. Brian W. Dippie, "Charles M. Russell, Cowboy Culture, and the Canadian Connection," in Simon Evans, Sarah Carter, and Bill Yeo, eds., *Cowboys, Ranchers, and the Cattle Business: Cross-Border Perspectives on Ranching History*

(Calgary: University of Calgary Press, 2000), 12–13; Warren M. Elofson, *Frontier Cattle Ranching in the Land and Times of Charlie Russell* (Seattle: University of Washington Press, 2004), 137; Granville Stuart, *Forty Years on the Frontier* (Cleveland, OH: Arthur H. Clark, 1925), 236; John Taliaferro, *Charles M. Russell: The Life and Legend of America's Cowboy Artist* (1996; repr., Norman: University of Oklahoma Press, 2003), 66.

2. "A Cow-Boy Artist," *Independent Record* (Helena, MT), May 1, 1887; "City and County," *Billings (MT) Gazette*, May 12, 1887; "A Diamond in the Rough," *Helena (MT) Weekly Herald*, May 26, 1887. A chinook is a warm wind, formally known as a föhn wind, that blows on the downwind, eastern side of the Rocky Mountains and brings unseasonably warm temperatures in winter that last for a few days or less.

3. William H. Goetzmann and William N. Goetzmann, *The West of the Imagination* (Norman: University of Oklahoma Press, 2009), 265–66.

4. John C. Ewers, *The Horse in Blackfoot Indian Culture, with Comparative Material from Other Western Tribes* (Washington, DC: Smithsonian Institution Bureau of Ethnology, 1955), 321; Karen R. Merrill, "Domesticated Bliss: Ranchers and Their Animals," in Matthew Basso, Laura McCall, and Dee Garceau, eds., *Across the Great Divide: Cultures of Manhood in the American West* (New York: Routledge, 2001), 170–71; Michael D. Wise, *Producing Predators: Wolves, Work, and Conquest in the Northern Rockies* (Lincoln: University of Nebraska Press, 2016), 74–75.

5. Dippie, "Charles M. Russell," 12–14. Examples of economic and land management interpretation of the Big Die-Up are too numerous to list; see, for instance, Elofson, *Frontier Cattle Ranching*, 139–43; Robert S. Fletcher, "That Hard Winter in Montana, 1886–1887," *Agricultural History* 4, no. 4 (October 1930): 126–28; Taliaferro, *Charles M. Russell*, 62–65.

6. Similar numbers of sheep also died, an important story that cannot be told here. Andrew Gulliford, *The Woolly West: Colorado's Hidden History of Sheepscapes* (College Station: Texas A&M University Press, 2018).

7. Joshua Specht, *Red Meat Republic: A Hoof-to-Table History of How Beef Changed America* (Princeton, NJ: Princeton University Press, 2019), 9, 70, 74, 107–8, 110.

8. Specht, *Red Meat Republic*, 110–12; Nathan F. Sayer, *The Politics of Scale: A History of Range Science* (Chicago: University of Chicago Press, 2017), 7, 9, 27–29.

9. Andrew A. Robichaud, *Animal City: The Domestication of America* (Cambridge, MA: Harvard University Press, 2019), 7.

10. Criollo cattle consist of various breeds all descended from Spanish stock

that live throughout the Americas as far north as the US sunbelt states. The group includes the famous Texas longhorns and dozens of other short-horn beef breeds native to South and Central America. Criollo are distinct from the heavier European short-horn cattle breeds imported to North America beginning in the seventeenth century, such as Hereford or Angus.

11. Maria-Aparecida Lopes and Paolo Riguzzi, "Borders, Trade, and Politics: Exchange between the United States and Mexican Cattle Industries, 1870–1947," *Hispanic American Historical Review* 92, no. 4 (October 2012): 603–35.

12. J. Frank Dobie, *The Longhorns* (1941; repr., Austin: University of Texas Press, 1982), 37.

13. Larry D. Christiansen, "The Extinction of Wild Cattle in Southern Arizona," *Journal of Arizona History* 29, no. 1 (Spring 1988): 89–100; see also Douglas Branch, *The Cowboy and His Interpreters* (New York: D. Appleton, 1926), 13; Frank J. Dobie, "The First Cattle in Texas and the Southwest: Progenitors of the Longhorns," *Southwestern Historical Quarterly* 42, no. 3 (January 1939): 171–97, 186–89, 192.

14. Christiansen, "Extinction of Wild Cattle," 93–94.

15. Kristin Hoganson, "Meat in the Middle: Converging Borderlands in the U.S. Midwest, 1865–1900," *Journal of American History* 98, no. 4 (March 2012): 1028–31, 1042–44; Walter Baron von Richthofen, *Cattle-Raising on the Plains of North America* (New York: D. Appleton, 1885), 58.

16. Dobie, *The Longhorns*, 18.

17. Although this study addresses dairy cattle, Moran and Doyle's explanation of feral cattle social organization certainly holds for beef cattle of various kinds. John Moran and Rebecca Doyle, *Cow Talk: Understanding Dairy Cow Behavior* (Clayton, Australia: CSIRO, 2018), 41–43.

18. Ramon F. Adams, *The Old-Time Cowhand* (New York: MacMillan, 1961), 154; Dobie, *The Longhorns*, 11; Marie-France Bouissou, Alain Boissy, Pierre Le Neindre, and Isabelle Veissier, "The Social Behavior of Cattle," in L. J. Keeling and H. W. Gonyou, eds., *Social Behaviour in Farm Animals* (Wallingford, UK: CABI, 2001), 114.

19. Walter Prescott Webb, *The Great Plains* (New York: Grosset and Dunlap, 1931), 242.

20. Fay E. Ward, *The Cowboy at Work* (1958; repr., Mineola, NY: Dover, 2003), 62.

21. Lily N. Edwards-Callaway, "Sense and Sensibility: Understanding How Cattle Perceive Our Collective World," in Terry Engle, Donald J. Klingborg, and Bernard E. Rollin, eds., *The Welfare of Cattle* (Boca Raton, FL: CRC Press, 2019), 64–65.

22. Specht, *Red Meat Republic*, 6, 99.

23. David Igler, *Industrial Cowboys: Miller and Lux and the Transformation of the Far West, 1850–1920* (Berkeley: University of California Press, 1990), 123-24.

24. Specht, *Red Meat Republic*, 99–100.

25. Collins was an Iowan who moved to a ranch in Oklahoma (then designated "Indian Territory") in the mid-1880s, where, as an adolescent, he worked as a novice cowboy with a particular interest in cattle behavior and intelligence. The cattle he knew in Oklahoma were the same kind of criollo cattle herded into the northwestern Great Plains. Hubert E. Collins, *Warpath and Cattle Trail* (New York: William Morrow, 1928), 147, 153. See also Dobie, *The Longhorns*, 17–18.

26. Hubert E. Collins, *Warpath and Cattle Trail (New York: William Morrow, 1928)*, 150. See also John R. Barrows, *Ubet* (Caldwell, ID: Caxton Printers, 1936), 57.

27. Collins, *Warpath and Cattle Trail*, 147.

28. Dobie, *The Longhorns*, 176–83. Cattle can smell the hormones released into urine by other cattle who are frightened and, as a prey species, will act accordingly to get to safety. Moran and Doyle, *Cow Talk*, 42–43.

29. Collins, *Warpath and Cattle Trail*, 151; see also Barrows, *Ubet*, 192; Branch, *Cowboy and His Interpreters*, 74. Today there is still only limited academic research explaining the nuances of communication among cattle, although some research demonstrates (for academics) that, for instance, they can identify one another as individuals, especially in cow and calf pairs. Presumably, the lack of research into the specifics of cattle vocalizations is due to the fact that most cattle research is funded and structured to serve beef and dairy production, not any understanding of cattle for their own sake, unlike studies of wild species such as coyotes or chimpanzees, for instance. So, generally speaking, the research questions that many scientists explore are funded only if they can be justified as increasing agricultural production or human worker safety. See, for instance, Gary M. Landsberg, "Social Behavior of Cattle," in *Merck Manual Veterinary Manual*, https://www.merckvetmanual.com/behavior/normal-social-behavior-and-behavioral-problems-of-domestic-animals/social-behavior-of-cattle; J. M. Watts and J. M. Stookey, "Vocal Behaviour in Cattle: The Animal's Commentary on Its Biological Processes and Welfare," *Applied Animal Behaviour Science* 67 (2000): 16.

30. Dobie, *The Longhorns*, 177.

31. Collins, *Warpath and Cattle Trail*, 147–48; Rollins, *Cowboy*, 179.

32. Sayer, *Politics of Scale*, 10, 12.

33. Edward Charles Abbott, *We Pointed Them North: Recollections of a Cowpuncher* (1939; repr., Norman: University of Oklahoma Press, 1955), 15.

34. Andrew C. Isenberg, *The Destruction of the Bison: An Environmental History,*

1750–1920 (2000; repr., Cambridge: Cambridge University Press, 2020), 9–10.

35. Adams, *The Old-Time Cowhand*, 160; Emerson Hough, *The Story of the Cowboy* (New York: Grosset & Dunlap, 1897), 346; Philip Ashton Rollins, *The Cowboy: His Characteristics, His Equipment, and His Part in the Development of the West* (New York: Charles Scribner's Sons, 1922), 225.

36. Barrows, *Ubet*, 140; Robert H. Fletcher, *Free Grass to Fences: The Montana Cattle Range Story* (New York: University Publishers Incorporated/Historical Society of Montana, 1960), 15–27, 39; Pekka Hämälainen, *Lakota Empire: A New History of Indigenous Power* (New Haven, CT: Yale University Press, 2019), 296, 375–76.

37. Elofson, *Frontier Cattle Ranching*, 140.

38. Umeek / E. Richard Atleo, *Principles of Tsawalk: An Indigenous Approach to Global* Crisis (Vancouver: University of British Columbia Press, 2012), 37. See also Richard White, "Animals and Enterprise," in Clyde A. Milner II, Carol A. O'Connor, and Martha Sandweiss, eds., *The Oxford History of the American West* (Oxford: Oxford University Press, 1994), 237–38.

39. Branch, *The Cowboy and His Interpreters*, 107.

40. Some key titles here were Joseph G. McCoy's *Historic Sketches of the Cattle Trade of the West and Southwest* (1874), James S. Brisbin's *The Beef Bonanza; Or, How to Get Rich on the Plains* (1881), and Walter Baron von Richthofen's *Cattle-Raising on the Plains of North America*, plus countless boostering newspaper and magazine items. See also Joseph Kinsey Howard, *Montana: High, Wide, and Handsome* (1943; repr., Lincoln: University of Nebraska Press, 2003), 138–39.

41. Von Richthofen, *Cattle-Raising on the Plains*, 56.

42. Howard, *Montana*, 149.

43. Rollins, *The Cowboy*, 39.

44. Specht, *Red Meat Republic*, 106.

45. Sayer, *Politics of Scale*, 11, 28–29.

46. Michel T. Kohl, Paul R. Krausman, Kyran Kunkel, and David M. Williams, "Bison versus Cattle: Are They Ecologically Synonymous?," *Rangeland Ecology Management* 66 (November 2013): 721–31.

47. There was a series of winters with terrible storms in 1880–81 in Montana, in 1885 and 1886 on the central Plains, and in 1886–87 across the whole of the Great Plains. Barrows, *Ubet*, 103; Fletcher, *Free Grass to Fences*, 43–51; John T. Schlebecker, *Cattle Raising on the Plains: 1900–1961* (Lincoln: University of Nebraska Press, 1963), 7; Frank Wilkeson, "Cattle-Raising on the Plains," *Harper's New Monthly Magazine* 72, no. 431 (April 1886): 793. In Alberta, overstocking plus a

deadly winter created another die-up in 1906. Elofson, *Frontier Cattle Ranching*, 140.

48. William Cronon, *Nature's Metropolis: Chicago and the Great West* (New York: W. W. Norton, 1991), 220; Elofson, *Frontier Cattle Ranching*, 140; Specht, *Red Meat Republic*, 76–80.

49. Specht, *Red Meat Republic*, 81.

50. Wilkeson, "Cattle-Raising on the Plains," 793.

51. Abbott, *We Pointed Them North*, 174.

52. Sara Dant, *Losing Eden: An Environmental History of the American West* (Somerset, UK: Wiley-Blackwell, 2016), 92.

53. "Stock Department," *River Press* (Fort Benton, MT), December 22, 1886.

54. Peter Iverson, *When Indians Became Cowboys: Native Peoples and Cattle Ranching in the American West* (Norman: University of Oklahoma Press, 1994), 46; James N. Leiker and Ramon Powers, *The Northern Cheyenne Exodus in History and Memory* (Norman: University of Oklahoma Press, 2011), 45.

55. *Billings Gazette,* quoted in "Clearing the Reservation," *River Press* (Fort Benton, MT), December 22, 1886. Regarding fines for directing one's cattle onto reservation lands, see, for instance, "City and County," *Billings Gazette*, October 13, 1886.

56. Abbott, *We Pointed Them North*, 174–77.

57. Barrows, *Ubet*, 65.

58. Terry G. Jordan, *North American Cattle Frontiers: Origins, Diffusion, and Differentiation* (Albuquerque: University of New Mexico Press, 1993), 210–40.

59. Von Richthofen, *Cattle-Raising on the Plains,* 57. See also James S. Brisbin, *Beef Bonanza; Or, How to Get Rich on the Plains* (Philadelphia: J. B. Lippincott, 1881), 61.

60. Dobie, *The Longhorns*, 201.

61. Stuart, *Forty Years on the Frontier*, 236.

62. See for instance, Abbott, *We Pointed Them North*, 38; Adams, *Old-Time Cowhand*, 219.

63. Hough, *Story of the Cowboy*, 346–47.

64. Abbott, *We Pointed Them North*, 175–76.

65. Adams, *Old-Time Cowhand*, 219. Dry murrain was a contemporary term for a deadly condition caused when an animal ingested a large amount of indigestible material with insufficient water, such that the material formed a hard, dry, solid mass in the third stomach. "The symptoms of the disease are as follows: Quickened breathing, dry muzzle, dull eyes, moaning; animal is found by itself generally, lying with its head to one side; eyes glaring and evincing great pain. Death may ensue in

six or eight hours, or the animal may linger for several days." "Dry Murrain," *Fort Collins Courier*, January 24, 1884.

66. Ewers, *The Horse in Blackfoot Indian Culture*, 45.

67. "John Stands-In-Timber Interview—Biographical Note," Archives of the West, http://archiveswest.orbiscascade.org/ark:/80444/xv72369.

68. John Stands in Timber, *Cheyenne Memories*, ed. Margot Liberty (New Haven, CT: Yale University Press, 1967), 255.

69. Branch, *Cowboy and His Interpreters*, 117–18; Hough, *Story of the Cowboy*, 184–85; Rollins, *The Cowboy*, 226.

70. Adams, *Old-Time Cowhand*, 218.

71. Elofson, *Cattle Ranching Frontiers*, 101.

72. Adams, *Old-Time Cowhand*, 219.

73. Barrows, *Ubet*, 109.

74. See, for instance, Joseph Ferdinand Boissard de Boisdenier, *Épisode de la retraite de Russie* (1835), https://fr.wikipedia.org/wiki/Joseph_Ferdinand_Boissard_de_Boisdenier; and Adolph Northen, *Napoleon's Retreat from Russia* (1851), https://en.wikipedia.org/wiki/Adolph_Northen.

75. "City and County," *Billings Gazette*, February 2, 1887.

76. "The Stock Outlook," *Billings Gazette*, February 3, 1887.

77. Hough, *Story of the Cowboy*, 188.

78. "The Stock Outlook," *River Press* (Fort Benton, MT), February 16, 1887.

79. "The Stock Outlook," *River Press* (Fort Benton, MT), February 16, 1887; "Condition of the Cattle in Fergus County," *River Press* (Fort Benton, MT), February 16, 1887.

80. Stuart, *Forty Years on the Frontier*, 237.

81. John Clay, *My Life on the Range* (Chicago: John Clay, 1924), 208.

82. Webb, *The Great Plains*, 240.

83. On twenty-two-hour laws and other attempted regulation of cattle welfare in rail transit, see Diane L. Beers, *For the Prevention of Cruelty: The History and Legacy of Animal Rights Activism in the United States* (Athens: Ohio University Press, 2006), 70–72; Ernest Freeberg, *A Traitor to His Species: Henry Bergh and the Birth of the Animal Rights Movement* (New York: Basic Books, 2020), 120–39.

84. Freeberg, *Traitor to His Species*, 150–51; Howard, *Montana*, 162–64.

85. White, "Animals and Enterprise," 266.

86. Webb, *The Great Plains*, 243.

87. Merrill, "Domesticated Bliss," 173.

FIG. 4.1. Stephen Nelson Leek, *Elk Herd Eating Hay in Snow on the Leek Ranch, Jackson Hole, WY*, no date, published in *Zoological Society Bulletin* in May 1914. Photographic print, original version glass plate negative, 16.5 × 21.6 cm. S. N. Leek Papers, University of Wyoming, American Heritage Center.

VANESSA BATEMAN

four Animal Photography and the "Elk Problem" in Modern Wyoming

In the May 1914 issue of the New York Zoological Society's bulletin two photographs document the fluctuating population of the Jackson Hole elk herd and offer a glimpse of the larger conflicts between man and animal brought on by the arrival of Euro-American settlers in this Wyoming valley. The first image resembles the aftermath of a battle: a group of dead elk are strewn across a landscape. In the foreground their bodies are emaciated, intertwined, and distorted, seeming to have been there for a long time, perhaps since the last snowfall, which is melting on the hill in the background. In contrast, the second photograph depicts a large herd of live, healthy elk feasting on hay in a winter field (fig. 4.1). These elk are even more numerous than the first picture, receding far into the background space and blending into the landscape. The first photograph was taken five years earlier than the second, in 1909; viewed chronologically, the pictorial narrative of dead-to-alive is explained by accompanying captions. The first reads, "Five years ago, when elk were starving. Photographed in Jackson Hole, Wyoming by S. N. Leek before Congress took hold to save the elk." The second: "Elk herd today in Jackson Hole. They have been fed all winter by the Government and will seek their grazing ground in the hills as soon as the snow goes off. There are 887 elk in this picture and behind the camera are many more. Photographed by S. N. Leek. February 1914."[1] As a visual record of animal history, this diptych represents the elk as both a vulnerable and a protected species, animals who increasingly lost their self-sufficiency and ended up reliant on the interventions and control of Euro-American settlers who had caused the crisis in the elk's livelihood in the first place.

This chapter discusses the intersections among animal photography, hunting, and conservation in Jackson Hole, Wyoming, and the issues surrounding the "elk problem"—that is, the mass starvation and death of thousands of elk due to Euro-American settlement between 1889 and the establishment of the National Elk Refuge in 1912. Publicizing the demise of the elk while also popularizing the area as a wildlife tourist destination, homesteader Stephen Nelson Leek (1858–1943) became the leading advocate for the preservation of the elk. The national circulation of Leek's photographs gave pictorial access to wildlife of the American West in an otherwise remote part of the country; even in 1910 Jackson Hole was 120 miles from the nearest railway station in Idaho and accessible in the winter only by snowshoe through Teton Pass. Just as images were used to help establish and promote the first national parks in the nineteenth century, photographs of animals in the wild were published to sway public and congressional opinion about their protection. This chapter reveals how the simultaneous arrival of cameras in the region and the increasing issues facing elk resulted less in an environmental crisis than an opportunity to colonize the animals through federally funded feeding programs and force them into restricted territories. Wildlife preservation served not only to advocate for the well-being of animals and their habitat but also to gain control over animals for economic and political purposes.

"Visual imagery," writes Alan C. Braddock, "far from being transparent or of negligible importance, played a central role in mediating environmental attitudes and politics in late-nineteenth-century America."[2] In the tradition of promoting the American West through photography and painting, animals and the landscape were represented as independent from humanity—as unspoiled and idealized. When Euro-American settlers were depicted, they were portrayed as being at the mercy of nature. Painters like Charles M. Russell perpetuated the romanticization of the West through images of its ostensible dangers: predatory animals, extreme landscapes, severe weather, and Indigenous people. The camera diverges from other mediums, however, in that it was believed to be "a recorder of facts," a machine of "great scientific value, for it cannot lie."[3] Understood as an objective representation of the world, photography was capable of persuading viewers to support environmental conservation. Most famously, Carleton Watkins's photographs of Yosemite Valley represented the landscape as pristine and void of humanity—settler or Indigenous—and these images were used in 1864 when Congress passed and President Lincoln signed the first legislation to preserve land for the common good. Similarly, William Henry Jackson's photographs and Thomas Moran's paintings of the Yellowstone region during the Hayden Expedition, a federal survey in 1871, were used in the official report to lobby Congress to establish Yellowstone National Park in 1872. In fact, Congress purchased Moran's painting *The Grand Canyon of the Yellowstone* (1872), which was then widely popularized by chromolithograph reproductions—circulating conservationist ideals about the American wilderness as pristine and hiding the realities of violence, death, and destruction inflicted on Indigenous peoples, animals, and the environment.

Leek's photography diverged from traditional portrayals of nature by maintaining the copresence and codependence of white settlers and animals, and documenting the ongoing struggle to gain control of environmental problems, including pictures of dead and dying animals in the landscape. Beginning his photographic career in keeping with the emerging genre of "camera hunting" at the time, Leek's practice combined a hunter's visualization of elk while also documenting their existence, peril, and survival in relationship with white settlers. His pictures recorded what it was like to be an elk in Wyoming at the turn of the century, as animals confronted multiple obstacles, including spatial and ecological changes as well as encounters with poachers, hunters, and photographers on the landscape. By publicizing the demise of the elk to help them, while also popularizing Jackson Hole as a hunting destination, Leek's photographs reveal that Euro-American settlers wanted to protect elk both as wildlife of intrinsic value and as a natural resource to be exploited by people.

THE BEGINNING OF THE ELK PROBLEM

The history of Euro-American settlement in Jackson Hole is in keeping with the "story of frontier migration" William Cronon describes as "a long tale of people moving to frontier areas, seizing abundance, encountering scarcity, and remaking the land and themselves in the process."[4] Fifty miles south of Yellowstone National Park, Jackson Hole (originally called Jackson's Hole) is a large valley situated between the Teton and Gros Ventre mountain ranges in Wyoming. Teton County is home to a diversity of plants and a rich variety of wildlife, with over fifty species of mammals who inhibit the region part or most of the year. Surrounded by rugged mountains, the basin was inhabited by prehistoric people as early as 6500 to 7400 BCE, eventually belonging to the Eastern Shoshone nation, who occupied the valley during the summer months for ceremonial and hunting purposes.[5] With an average elevation of

6,800 feet (2,100 meters), its geographic remoteness, rough terrain, and climate led to the relatively late arrival of homesteaders at the end of the nineteenth century, when the rest of the western frontier had already reached the Pacific. The arrival of white settlers in Jackson Hole, beginning in 1884, altered the landscape, causing ecological changes that disrupted wildlife populations, migration patterns, habitats, and resources.

Significantly impacted were the herds of elk who migrated in the thousands from the Teton Mountains and the Yellowstone region to their winter ranges in the Red Desert and Green River Basin areas of southern Wyoming. The Rocky Mountain elk (*Cervus canadensis nelsoni*), a subspecies of elk found in the Rocky Mountains and the ranges of western North America, is one of the largest mammals in North America. These ungulates are also called wapiti, from the Shawnee and Cree word *waapiti*, meaning "white rump." Elk are social animals who herd together while traveling, and their seasonal migrations take place from higher altitudes in the summer, when their large matriarchal herds are dominated by one or more mature cows, to lower altitudes during the winter months. Today elk spend the spring, summer, and early autumn in the meadows of the Yellowstone and Grand Teton National Parks, before they subdivide into discrete herds, including the southern or Jackson herd, to winter outside the parks.[6]

Prior to Euro-American settlement, there is evidence that the southern herd used Jackson Hole only as a resting ground along their journey, but by the late 1880s the herd began shortening the migration route so that Jackson Hole became their winter destination, and this pattern has continued since that time.[7] However, Jackson Hole was not ideal, as ranching, farming, and development soon began to encroach on elk feeding grounds. Descending into the valley, elk faced numerous changes along their historic migration route: extensive irrigation, the replacement of sage-brush flats and meadows on which they foraged with fields of grain and hay, and newly built barbed-wire fences and roads. There was also new competition from domestic livestock who ranged the land that had previously supported elk and other animals. The changes brought to the landscape by animal agriculture demonstrate that white settlers were not individual immigrants but, as Alfred Crosby shows, were "part of a grunting, lowing, neighing, crowing, chirping, snarling, buzzing, self-replicating and world-altering avalanche."[8] Faced with these changes, elk were confined to Jackson Hole and the surrounding area in the winter and during particularly harsh winters were at the mercy of severe cold and heavy snow accumulation. Some years thousands of elk died of starvation, succumbing to what was known as the elk famine or "elk problem."

Some researchers argue that the loss of traditional forage due to farming and the introduction of hay, even when it was insufficient, created a reliance on the new food supplied by people.[9] While modest in size, the increasing cattle population in the valley coincided with the emergence of the elk problem. According to the Wyoming Stock Growers Association, ranches in Jackson Hole steadily increased; in 1901, there were 29 ranchers and 1,540 cattle, and by 1910 there were 61 ranchers and 10,919 cattle (although these numbers are considered low because this does not account for cattle raised by nonmembers).[10] Unlike the Great Plains, the size of ranches was limited in Jackson Hole because of the short grazing season, and herd size was determined by the amount of land needed to cultivate hay in the winter.[11] It is in the periphery of ranches, or on them, that many elk died of starvation.

Another contribution to the elk problem was the situation in the Jackson herd's summer ranges. When Yellowstone National Park was set aside as a "pleasuring-ground for the benefit and enjoyment of people," little was initially done to protect its wildlife.[12] Poaching was so rampant in the park that an estimated four thousand elk were hunted every year from the 1870s until the US Army took over in 1886. According to historian Paul Schullery, around the time Yellowstone was established, hunting had caused an "ecological holocaust."[13] It was not until 1894, when the Yellowstone Game Protection Act was passed, that the situation changed. But these conservation efforts were counterproductive: by the late 1890s elk numbers increased so greatly that they exceeded the park's capacity to feed them, and by the 1920s and 1930s heavy grazing and browsing by elk was killing plants, shrubs, and trees, and causing soil erosion.[14]

In the hierarchy of wildlife, elk were considered "the most valuable livestock in Wyoming" and "the stateliest deer in the world," and requiring protection from predatory animals.[15] In contrast, bounties were placed on such animals as mountain lions, wolves, coyotes, wolverines, bobcats, and bears because they were viewed as threats to game and livestock, species of economic and cultural value. At least seven thousand elk were slaughtered between 1875 and 1877 for "their hides, or perhaps their carcasses, which were stripped and poisoned for bear, wolf, or wolverine bait."[16] Bounties were encouraged and paid for by private landowners, rancher associations, the territory and state, and eventually by the federal government, demonstrating how wildlife would become intensely human managed. Perhaps the most vilified species in Wyoming was the gray wolf, which was increasingly perceived as an economic threat to both ranching and game animals. Between 1895 and 1927 over 36,000 wolves were killed by bounty hunters and government trappers, contributing to their near extinction from Wyoming by 1930.[17]

In what was once regarded as a landscape of "endless abundance," the elk problem revealed how the frontier was, as Cronon describes, a landscape of "scarcity, so fragile in the face of human destructiveness that only careful management could ensure survival."[18] Outside Wyoming, elk had already succumbed to the irreversible impacts of Euro-American settlement. In the eastern United States and southern Canada the eastern elk (*Cervus canadensis canadensis*) was declared an extinct subspecies in 1880, around the same time the Merriam's elk (*Cervus canadensis merriami*) was made extinct in the southwestern United States due to overhunting and cattle grazing.[19] At stake was not just the future of the elk as a species but the livelihood of the new community, which relied on the growing recreational tourism economy, which entailed guided fishing, trophy hunting, and later, wildlife photography. There was the economic value of elk as a material good, in the form of live animals, meat, hides, antlers, and eyeteeth, all significant exports from the state. Therefore, saving the elk from human impact would also ensure there would be enough animals of value. At the same time, many farmers considered (and still consider) elk to be pests since they can damage crops and fences, eat livestock feed, and possibly spread disease. With little support or infrastructure from the territory and state to deal with these issues, residents of Jackson Hole found ways to preserve wildlife on their own terms.

S. N. LEEK, "FATHER OF THE ELK"

Sportsmen conceptualized elk as animals to be hunted (by camera or rifle) in a manner of fair chase and to be protected from the inappropriate deaths they increasingly faced in the forms of "poaching" and starvation. But how do we find the animal experience within these human sources? Bearing witness to the spatial and ecological changes, as well as the multiple ways human power structures and ideologies affected the lives of the local wildlife, the photographs allow us to piece together what it was like to be an elk in Wyoming at the turn of the century. The archive of Leek's photographs, which spans over fifty years (1890–1943), contains thousands of photographs of wildlife in the Jackson Hole region, with a particular focus on elk—both dead and alive—exhibiting a visual narrative of the animals as a revered species.[20] Pictured in situ, many elk gaze directly at Leek's camera, as if engaging for a brief moment with their human observer. Seen through Leek's eyes and lens, elk were noble, primordial, even ideal creatures that stood in for the landscape itself. Widely published between 1890 and 1920, Leek's pictures and writing appeared in sporting magazines, newspapers, and scientific books, as well as federal and state publications. In the beginning, he sold and published photographs as a way to promote himself as a hunting guide for sportsmen from the East. In the late 1890s until the 1910s, sporting magazines such as *Recreation* frequently printed letters Leek wrote with updates on big game hunting in Wyoming, illustrating their pages with photographs he mailed in. At the same time, Leek began documenting the aftermath of poaching, the starving and dying elk on his ranch, and the efforts to save them through supplemental feeding, distributing the photos and urgent letters to sporting journals to call attention to these issues.

"HUNT WILD ANIMALS WITH A CAMERA"

With the rapid development of the American conservation movement in the nineteenth century, growing concern for protecting land extended to an effort to preserve wildlife, especially with the anthropogenic near extinction of bison by the 1880s and the extinction of the passenger pigeon by 1901. Recreational hunters in particular wanted to protect individual big game species that were facing issues of habitat destruction and overhunting, organizing their efforts through private sportsmen's clubs and the sporting press to pressure local and national governments to create hunting regulations.[21] Their goal was not only to save game animals from extinction but to ensure sport hunting in the future. Coinciding with these issues were the technological advancements made in photography, which had the potential to support wildlife conservation efforts.

It was not until the late nineteenth century, which brought faster film and compact cameras, that photographing animals in their natural habitats became possible. Prior to these developments even the slightest movement of subject or camera caused blurring of the image, which explains why some of the first photographs of animals in their natural habitat were actually of taxidermy. The first photographs of living animals in the wild were taken in the 1890s by hunters who turned to photography as a new technology to accompany or replace their weapons. Classified as "camera hunting," this early form of animal photography emerged when hunters employed methods similar to those used to shoot animals with a firearm.[22] When Leek received a camera from Eastman around 1891, he adapted hunting skills analogous to those that would produce successful photographs of animals in the wild. They include a working knowledge of animals and their habitat, the use of specific tools, precision in aiming at "prey," and familiarity with shooting technology, whether weapon or camera. Furthermore, photographing animals in the wild was believed to be equal to, if not more difficult than, hunting itself since the technological and logistical limits of early photography made it extremely challenging. Early articles on camera hunting, including those by Leek, describe in detail how to produce a successful shot. Consider, for instance, the quick shutter speed and adequate sunlight needed to capture the unpredictable movement of animals in the wild while navigating wind, sun direction, and other on-site variables. Camera hunting required the use of other living animals, too, such as horses or mules, to bear the burden of packing in and out the excess weight of fragile glass-plate negatives, equipment, and chemicals, as well as kills, whether it be meat for camp or trophy heads.

Increasingly, camera hunting was elevated over hunting as a way to preserve wildlife—a position advocated, though not practiced, by Theodore Roosevelt, who observed, "More and more, as it becomes necessary to preserve the game, let us hope that the camera will largely supplant the rifle."[23] Stricter hunting restrictions across the United States coincided with the rise of new photographic technologies that made it easier and more affordable to become an amateur camera hunter. As a result, this new hobby was marketed in popular publications such as *Recreation*, *Forest and Stream,* and *Outdoor Life* to sportsmen who were looking for alternative ways to hunt when the season was over, with advertisements for camera equipment and photographic paper appearing alongside ads for ammunition and hunting lodges. Kodak explicitly suggested that the camera could replace the gun, using slogans such as "There are no game laws for those who hunt with a Kodak" or advising "There are no game laws—and no accidents—for those who hunt with a Kodak" and "The rod or the gun may be left out, but no nature lover omits a Kodak from his camp outfit."[24]

With a similar sentiment, in 1917 two of Leek's photographs of elk herds in the landscape were included in the brochure *The National Parks Portfolio* alongside an article that suggested, "Hunt wild animals with a camera."[25] Since hunting was restricted within park boundaries, encouraging visitors to take up photography contributed to the wider ethos at stake, akin to the conservationist motto "Take only pictures, leave only footprints."[26] The implication here is that hunting with a camera was less intrusive than hunting with a gun; yet, as Matthew Brower argues, "This rhetoric positions nature photography as maintaining a separation between humans and nature" and "assures us that photography keeps us at an appropriate distance from nature."[27] Closer examination of these images and their narratives, however, reveals how intrusive the practice could be. Although claiming to portray animals ostensibly free of human influence, these modes of photography were often artificial and required intrusive hunting tactics. As Brower demonstrates in *Developing Animals*, the widespread use of trip wires and bait and the practice of "jacklighting" (a hunting technique, now illegal, that uses fire or light to attract and stun animals) confirm that camera hunting was by no means a passive activity.[28] Some photographers used dogs to tree or chase animals in order to take their photographs, only to kill them afterward.[29] A Wyoming game law introduced in 1906 illustrates how animal photography was equated with hunting, requiring photographers to obtain a permit from the state game warden as well as employ a special assistant warden to supervise the expedition at the salary of $3 per day.[30] From written accounts, however, it appears Leek was less intrusive in his tactics of photographing living animals, wearing white clothes and painting his camera white to blend with the snow and later feeding hay to elk on his ranch as a convenient way to lure the animals closer to the camera.

By picturing human-animal encounters, photographs of animals by way of "camera hunting" lacked "the separation of human and animal that is central to wildlife photography."[31] For some, hunting with a camera was never a matter of replacing the rifle. In fact, obtaining a picture before and after a kill was common, as seen throughout Leek's archive, including *The Elk of Jackson's Hole, Wyoming, Their History, Home and Habits*.[32] The photographs in this self-published book show a progression of human-animal encounters from the hunter/photographer's perspective. As the title clearly implies, the reader is given a history of the elk, but only from a hunter's perspective. We see, first, a close-up picture of a large bull who faces the camera, followed by an image of an unnamed hunter aiming at a herd of elk in the distance, then the same hunter posed with his "prize head," and the book concludes with several pictures of large herds of elk in the landscape. Leek's pictorial narrative promotes his marksmanship as a camera hunter and at the same time demonstrates how pictures provide proof of the abundance of game animals in the region. One could, in fact, "prove it—with a Kodak."

"INDIAN HUNTING RIGHTS"

As "the father of the elk," Leek's authority on elk was a white settler's vision of a sustainable relationship between humans and animals, one which, through his direct participation in racially targeted conflict, perpetuated an intolerance of Indigenous hunting. This included the belief that Indigenous tribes were intruders in the valley and contributors to the population decline of game animals, even though these forms of subsistence hunting had been sustainable for thousands of years. The arrival of Euro-American settlers in the mid-nineteenth century encroached on Indigenous territory, created competition for resources, and caused damage to the habitat, altering the behavior of the game on which their livelihood depended. Often ignoring game laws themselves, as Karl Jacoby observes, the residents of Jackson Hole viewed subsistence hunting by settlers "as a natural right" and "rarely opposed poaching done for necessities such as meat, hides, or tallow."[33] In contrast, much like elsewhere in the country, off-reservation hunting by Native Americans was considered poaching and a threat to the livelihood of game animals and white hunting guides. People of Jackson Hole were specifically intolerant of the annual hunting practices of the Shoshone and Bannock because they did not have to follow the same game laws as non-Indigenous peoples since they were guaranteed by the 1868 Treaty of Fort Bridger "the right to hunt on the unoccupied lands of the United States so long as game may be found thereon."[34]

Conflicts regarding hunting rights increased throughout Wyoming, Montana, and Idaho in the 1890s, culminating in the Bannock Uprising, or Bannock War of 1895, which began when twenty-seven armed white men from Jackson Hole confronted a Bannock family who had traveled from Fort Hall Reservation to hunt elk, a traditional food source, to the south of Yellowstone.[35] Among the mob was Leek, who participated in arresting the families for "wantonly killing game," a confrontation that became violent when Bannocks attempting to escape were shot at, resulting in one death and several injuries.[36] The unprovoked violence led to a national public outcry, a federal inquiry, and eventually the US Supreme Court case *Ward v. Racehorse*, which declared that the Treaty of Fort Bridger was superseded by Wyoming law and required Indigenous persons to follow state hunting regulations off reservation.[37] Increasingly, access to a dwindling population of game animals was a privilege of the white settlers who had caused the animal population to be in a crisis in the first place. The Bannock Uprising reveals how the anthropogenic environmental change that came with colonization had not only put new pressure on elk populations but it also exposed the political inequalities between white settlers and Indigenous peoples.

In the months following the conflict, *Forest and Stream* published several articles on the matter. While most of these shared the belief of Jackson Hole residents that Indigenous hunting practices were equal to that of poachers and market hunters, on several occasions the editor condemned the conflict as murder, pointing out that hunting parties from the East were equally, if not more, detrimental to elk populations but had not met the same reaction.[38] Updating readers on the appeal taken to the Supreme Court, the article "Indian Hunting Rights" featured three of Leek's photographs, the first time he was published in *Forest and Stream*. Leek had submitted these pictures as a direct response to the recent conflict, writing, "Find inclosed [*sic*] some photos, taken by myself last winter, of elk on their winter range in Jackson's Hole. . . . Can not the sportsmen of the East with their influence help us, the people of Jackson's Hole, to save these noble animals from all being slaughtered?"[39] His position on the issue of Indigenous hunters perceived as "poachers" was not unusual for the time; in fact, Leek was explicit in letters to various publications on the matter, blaming a decline in the elk population on the presence of Indigenous peoples on the land and boasting in 1896 how "one bunch of Indians have been arrested and fined . . . [and] now languish in jail," perhaps the victims of the Bannock Uprising.[40]

The inclusion of his pictures is odd given the condemnation of the conflict by the journal and Leek's direct participation in it. Nonetheless, when published alongside the text, these animal photographs came to represent power and control over hunting rights and reveal how the relationship between humans and animals was understood from a settler-colonial viewpoint. Like most of Leek's early work, the photographs published in *Field and Stream* are taken from a distance, groups of silhouettes on the snowy landscape that create the illusion that we, the viewers, are looking at them through the barrel of a gun. From the lens of a white settler, Leek's camera-hunting images were reframed in the context of the conflict with Indigenous peoples, but at the same time, they reached the same sports community the photographs were initially intended for.

In the context of a sports publication, where hunting was pursued as a form of leisure, subsistence hunting lacked the same cultural capital as sports hunting. Between the nineteenth and twentieth centuries there was a signal change in how hunting was practiced and understood in the United States and how it represented a cultural identity. In the wake of the Civil War, when subsistence hunting declined, sport hunting grew in popularity among middle- and upper-class men. By the latter half of the nineteenth century the utilitarian function of the hunt was replaced by its value as a sport, serving pleasure rather than subsistence. This emphasis continued to define hunting during the twentieth century. As sociologist Jan Dizard observes, this shift furnished the grounds for the critique of hunting until recent times, as most critics of hunting accept its value as a utilitarian activity but not as recreation.[41] With a desire to connect with the tradition of elite British hunters, the ideal of the white sportsman appealed to the increasingly urban and educated middle- and upper-class American, or as Thorstein Veblen put it in 1899, the "predatory leisure class."[42] Sports hunting as a continuous tradition redolent of upper-class sensibility was reflected in the practices and values embraced by sportsmen who believed they had a moral way of thinking about hunting and nature, a code of conduct that was socially constructed and willingly followed. At the turn of the twentieth century outdoor recreational magazines such as *Field and Stream* were produced for the sport hunter, publishing articles on hunting practices and stories that often underlined sportsmanlike practices of hunting within limits and in the proper season. Read through these different modes of address to audience and purpose, it is impossible to read Leek's photographs as simply representations of animals in a landscape. Leek's images were ideologically inflected attempts to define Indigenous hunting practices, long sustainable, as forms of poaching and to pronounce them illegal.

TUSKING

In the late 1890s while elk were already facing colonial environmental crises, another threat they faced was a new form of illegal market hunting, called tusking, for the purpose of collecting their upper canine eyeteeth. During the winter months elk are vulnerable creatures. In large numbers they are highly visible against the white snow, and, unable to camouflage, they become easy targets. Just as hunters had killed buffalo for their tongue or hide, leaving the rest of the carcass where it fell, "tuskers"—transient men and opportunistic locals—hunted adult elk solely for their two "ivories." Tusking became such a problem that in 1904 Leek began documenting the aftermath of illegal market hunting in the valley, providing authentication on the issue for Wyoming's *Game Warden's Report* as well as articles in newspapers and magazines. Using Leek's photos, the *Los Angeles Times* described a common scenario in 1906: elk laboring through the snow with their young and suddenly sensing humans in the snow, the loud sounds of gunfire, and a mad scramble to escape.[43] Sometimes tuskers wore skis or snowshoes to chase the animals into deep snow, which held them in place while they were killed at close range.

Between 1890 and 1910 tusking was driven by the popularity of elk teeth as part of the official badge of membership (ironically) in the Benevolent and Protective Order of Elks, a fraternal organization founded in 1868 in New York. Mounted on watch fobs, cuff links, pins, and rings, elk teeth were in such high demand that imitation teeth were produced and sold as the cost of ivory increased. Initially it was not elk hunting that supplied the demand but rather the purchase of Indigenous garments from which teeth were removed to resell.[44] When demand for ivories increased among members of the Elks, in addition to collecting from animals they hunted, residents of Jackson Hole began to carry pliers with them in the winter in order to remove the tusks from dead elk they came across in the valley.[45] Tusk collecting from hunting was not detrimental until outsiders, primarily transient white men who traveled through the area, took notice of the lucrative opportunity and began illicitly killing elk for the sole purpose of collecting their teeth. Tuskers developed ways to avoid getting caught, including smuggling the easily hidden teeth and fastening hooves to the soles of their feet to hide their tracks in the winter and avoid game wardens.[46]

Already photographing issues surrounding the elk problem, Leek began documenting the work of the illegal hunters: otherwise healthy bulls found rotting in the landscape, their meat, hides, and antlers left for waste. In contrast to his other pictures of dead elk, these have bloated bodies, and their faces are often out of the camera's viewpoint, as if to give them some dignity in their death. Following Leek's lead, in 1907 Al Austen, a forest ranger and amateur photographer with the first telephoto lens in the valley, captured a crew of tuskers known as the Binkley gang in the process of killing elk and extracting tusks in Yellowstone, images of which were used as evidence in their trial.[47] A decade earlier, L. A. (Laton Alton) Huffman had documented the illegal hunting of buffalo in Montana, photographs reproduced as illustrations in William Temple Hornaday's report, *The Extermination of the American Bison*, produced for the Smithsonian Institution in 1889.[48] Visually, both Huffman and Leek's pictures are reminiscent of Timothy O'Sullivan's (1840–1882) documentation of the aftermath of the American Civil War, the camera bearing witness to anonymous bodies of soldiers strewn across the landscape.[49] Leek's photographs offer a different visualization of the animal experience—or the aftermath of it, at least. While Huffman's photographs depict an active engagement with the slain animals by including men skinning buffalo and horses in the background, Leek's pictures are quiet, as the animals blend into the ground and antlers resemble the tree branches above.

The disparity between hunting *with* a camera and documenting forms of hunting *on* camera indicates how human-animal relations were conceptualized by white settlers at the beginning of the twentieth century. Increasingly, photographs were taken during hunting trips to equally celebrate and authenticate successful kills. For those who followed the sportsmanship code of fair chase, often these pictures contained visual markers showing that hunting parties had behaved honorably, including displays of the legal limit of animals per hunter arranged side by side as a testament to an adherence to game laws.[50] Others created elaborate displays of their excessive and illegal kills. George O. Shields, editor of *Recreation*, took on the role of activist for wildlife conservation by publishing these photographs as a way to name and shame people he called "game hogs." The circulation of Leek's photographs of elk who had died from illegal hunting fit within this genre of what were considered inappropriate animal deaths as a way to not only raise concern for the animals but shame members of the Order of the Elks, many of whom were part of the social circles of those who supported environmental conservation. Due in part to Leek's advocacy, there was a national call to stop illegal hunting, including by President Theodore Roosevelt, who addressed the issue in his 1902 State of the Union Address, stating that "butchering off such a stately and beautiful creature as the elk for its antlers or tusks" was "against our national good."[51] Again, elk were made vulnerable by the arrival of Euro-Americans. From a self-sufficient population, the elk became reliant upon a small group of white settlers and the charity and advocacy of the wealthy from the East to protect them from the predators of the market.

THE ELK FAMINE UP CLOSE

The progression of the multiple crises facing the elk are spatially reflected in Leek's photographic archive from the 1890s until the formation of the National Elk Refuge in 1912. Over time, his photographs provide a clearer picture of the animal experience. That is to say, the combination of the elk problem with the development of photographic technologies influenced how Leek photographed animals. When Leek began documenting numerous elk dying of starvation in Jackson Hole, the resulting images provided a visual record of how animals experienced a loss of autonomy—too weak to care about a man with a camera—as they succumbed to the consequences of the colonial project. In these scenes we get a closer look at the animal experience, with the camera recording human-animal interactions in which both parties were trying to deal with a crisis: the elk with starvation and severe winter weather, and the ranchers with preventing elk from stealing their livestock's hay but also attempting to keep the elk from dying.

Homesteaders had only been in the valley for a few years when the first elk famine occurred in the winter of 1889–90, resulting in the death of an estimated twenty thousand elk out of a herd of fifty thousand. In 1900 Leek wrote to *Recreation* asking for advice on the issue, describing how elk "attacked the settlers' haystacks, already too small to meet the needs of the domestic animals." To which editor George O. Shields suggested they continue to feed the elk hay and grow "soft wood trees, such as willows and aspens, on which such animals browse," and which had been replaced by farm fields in the first place.[52] The elk famine did not seem to be an issue until almost two decades later, when three harsh winters between 1908 and 1911 caused thousands of elk to die. The mass die-off was aggravated by several factors: the population of elk had significantly increased partly due to conservation efforts, there had been several years of mild winters, their natural predators had been eradicated, and much of the winter range had been settled with farms, ranches, homes, and roads as the population of Jackson Hole increased.

In 1909 the snow depth in the valley reached three feet and was frozen on the surface, making it impossible for the winter grazing of an estimated twenty thousand elk. Too famished to fear humans, elk charged pastures and stables, with some reports of animals breaking through barn windows, and one individual eating the bristles of a broom. The community tried multiple solutions to prevent elk from raiding their feed supply, including creating new enclosure designs, installing barbed-wire fencing, deploying guard dogs, and firing guns to scare or kill elk. Some ranchmen pitched tents or made camp near their haystacks, spending the night during the most severe months of winter to protect their cattle's food supply.[53] Leek recalled the struggle between ranchers and the starving elk in *Colliers Weekly*: "These are trying times for the settler, who must sleep by his haystack during the cold winter night to save the scanty supply for his stock, and who may be awakened during the night to find elk standing over him feeding. When they are starving hardly any kind of fence will stop them; they roam at will over all the ranches, devouring everything in sight that can be eaten, even to willow brush one-half inch in diameter."[54]

In Yellowstone, elk were also wreaking havoc, repeatedly breaking down fences as they stole hay meant for antelope, deer, and mountain sheep.[55] In negotiating these interactions with the elk, citizens of Jackson Hole were faced with not only the need to ensure the winter survival of their livestock but general concern for the well-being and survival of animals dying at their doorsteps, animals who provided income in the summer from tourists.

There is a sense of urgency and hopelessness in the photographs of the elk famine, both from the animals and the photographer. Leek described such a feeling: "I took a photograph from my barn last evening, showing fifty elk, part of them within the corral, and at the time there were fifteen hundred head of elk within my field, all starving. I could feed a hundred or so, but did I commence I should soon have a thousand to feed, and I haven't the hay. . . . I feel almost like quitting and letting them all die, and have the worry over."[56]

Photographs from these winters record scenes of elk carcasses strewn across

FIG. 4.2. Stephen Nelson Leek, *Two Young Elk at Haystack Surrounded by Carcasses on the Leek Ranch, Jackson Hole, WY*, no date. Photographic print, original version glass plate negative, 16.5 × 21.6 cm. S. N. Leek Papers, American Heritage Center, University of Wyoming.

fields and bodies piled around hay corrals. *Two Young Elk at Haystack Surrounded by Carcasses on the Leek Ranch, Jackson Hole, WY* encapsulates the desperation facing the animals, with hay visible but unreachable (fig. 4.2). There is also a series of pictures in which an unidentified man stands beside a group of elk, posing in contemplation of the animals suffering around him, some of which are looking up at him. In another, a yearling is photographed caught on a fence, its legs slumped over on either side, too weak to move.

Written accounts found in local newspapers and sporting magazines also provide a glimpse of the animal experience. One rancher described a desperate scenario: "These noble animals have eaten bark off the trees, devoured almost all available food, such as brush and twigs, and have been breaking into farmers' haystacks by tearing down fences and causing other depredations in search for food."[57] Another account recalled, "Elk have been known to mount upon the fallen bodies of their companions and thus climb to the top of a thatched roof shed where they would voraciously devour the rotten hay or straw used as a roof covering."[58] During a single season Leek claimed he could "walk a half mile on the bodies of dead elk," and a ranchman "counted the bodies of sixteen hundred dead elk in the spring of 1909."[59] In the spring of 1910 it took "a team ten days . . . to haul away the dead elk on a 400-acre ranch in Jackson's Hole."[60] The stench of rotting bodies in the spring was enough to make several families temporarily relocate, and the waters of the rivers and streams became unsafe to drink.[61]

Some of Leek's photographs overtly illustrate the issues facing the elk and the ranchers. For example, several photographs are theatrically composed. In the foreground are dozens of bodies of elk who have died of starvation, some of which look like they have been dead longer than others. These animals are all situated beside a fenced and abundant haystack, the food source that would have saved them. In *Two Young Elk at Haystack* there are only two survivors in the scene, one looking toward and the other away from the viewer (or photographer), as if contemplating their present and future (fig. 4.2). Did Leek drag these carcasses and arrange them next to the haystack to emphasize the dire situation? He would not be the first to do so for the sake of pictorial narrative. Civil War photographer Alexander Gardner, for example, moved and posed corpses, and changed the weapon of a soldier to create more dramatic narratives in such photographs as *Sharpshooter's Last Sleep* (1863).[62] Even if Leek's pictures were orchestrated, they authenticated written accounts of the situation by visualizing the human and animal experience for outsiders to see.

Hoping to witness the elk problem firsthand in an effort to prove whether "such conditions had been overdrawn," Dillon Wallace, a writer for *Outing* magazine, visited Jackson Hole during the winter of 1911. With Leek providing the photography, Wallace produced a two-part article on the issues facing the elk and the settlers of the region. Using Leek as his guide, Wallace's report explains that the animal suffering was caused by environmental damage by white settlers. On the visibility of the death toll in the landscape, he wrote:

> I read the sickening story of the tragedy of the elk, written in bold characters on every field, on every hill and mountainside. . . . [A]ll along the trail from the Hoback to the Gros Ventre were scattered bones and tufts of hair of animals that had starved. Bark-stripped willows and quaking aspens and twigs

. . . gnawed down by famished animals in vain attempt to find sustenance in dead sticks told the story of misery and suffering. On the fields wherever I walked . . . were the bones of innumerable elk that had perished within two years.[63]

Paired with Wallace's account, Leek's pictures portray the dire reality for the elk, eliciting empathy from the reader, especially the images documenting the dying among the dead—scenes of carnage that again resemble images of the corpse-strewn battlefields of the Civil War almost half a century earlier.

Leek's distressing photographs were mailed to local and national newspapers and magazines, as well as official government publications in an effort to gain state and federal support for their conservation. For example, several of the same photographs that appeared in Wallace's articles in *Outing* also appeared in the *Chicago Daily Tribune* (1911), the state game warden's official reports (1911–13), and the US Department of Agriculture's special bulletin on the conditions of the elk in Wyoming (1911). In all of these formats, Leek's photographs served as authenticators that sought to evoke emotional responses from the viewer. Encountering images of dead animals was not unusual in the nineteenth or early twentieth century, when the hunting adventure genre grew in popularity. At the same time, when viewers saw Leek's documentation of suffering and dead elk in Wyoming, the near extinction of bison was fresh in the minds of the nation. In response, editors and readers responded in several publications with advice on how to solve the problem, including the suggestion that surplus elk be relocated to states that could "adopt them," as a way to rebuild populations (something that did occur in Yellowstone between the 1910s and the 1960s). Others proposed expanding the boundaries of Yellowstone, since the elk were already "wards of the United States" who were "housed and fed" with federal protection in the summer, only to be "turned out" in the winter.[64]

FEEDING THE ELK: DOCILE SUBJECTS

Another solution was to feed the elk the hay that they were desperately trying to eat, yet this went against the belief that "when wapiti get a taste of good hay, they seem to lose all idea of going back to the foothills to find feed for themselves."[65] In fact, there had been discussion of an amendment to the Wyoming game law in 1903 to make it a misdemeanor for "anyone to let his haystack go unprotected, because when the winter is very severe a taste of hay is almost sure to be followed by the starvation of some of the elk."[66] Yet rationing a portion of hay for the elk every winter at the rancher's own cost seemed the only solution, documented by Leek on his own ranch. In his pictures of elk on the ranch, hundreds of elk are in a feeding frenzy and the closely cropped scenes emphasize the size of herds and their ravenous hunger, reminding viewers of the chaos that had taken place and would again without continuation of the feeding program. In a letter to the governor in 1909 Leek wrote, "Enclosed herewith find pictures of elk eating hay in the Jackson Hole country, showing how quickly they accustom themselves to the new conditions. Think of wild animals becoming so tame in two short months that they will allow people to drive among them on the feedyard all the same as domestic cattle. Feeding them seems to be a complete success."[67] As a result, the Wyoming legislature appropriated $5,000 in 1909 to purchase hay the following year. In 1911 Congress appropriated $20,000 "to be made available immediately for feeding and protecting the elk in Jackson Hole," and the year after allocated $45,000 and guaranteed the creation of an elk refuge by merging 1,000 acres of public land with 1,760 acres of purchased land, which would be expanded in the following decades. This parcel of protected land formed the Elk Refuge in 1912, renamed the National Elk Refuge in 1940.

Taming the elk was considered beneficial to both settlers and animals: finally the elk had a place to source food during the winter, and settlers gained control of the persistent elk famine that had wreaked havoc on their ranches. To residents of Jackson Hole who had dealt with years of elk famine and the issues surrounding it, having a feeding program and allocated land for the elk away from their own properties was the best of both worlds. Elk became semidomesticated pets that settlers and the government had responsibility for, but also control over. The strict management of elk was withheld from Indigenous locals, so the elk could no longer contribute to Indigenous self-sufficiency either. This is not to equate the local Indigenous people with the elk, but there is a parallel of control. Like the Bannock and Shoshone, elk were forced onto restricted territories and made reliant on government-supplied food. In finding a solution to the elk problem, the animals were colonized, so that the issue became not so much an environmental crisis but an opportunity for white settlers who had learned to prefer a controlled relationship with wild animals.

In the years after the feeding program began, Leek documented the size of herds in the feedlot as a way to show the health of the herd as a result of human intervention. These photographs also provided naturalists with a better understanding of the population demographics of the elk, including William T. Hornaday, who included a photograph by Leek in his book *Our Vanishing Wildlife: Its Extermination and Preservation*.[68] While congratulating the people of Jackson Hole on preserving the elk, Hornaday noticed that as a result of hunting, poaching, and famine, there was "an alarming absence of mature bulls, indicating that now most of the breeding is done by immature males," concluding that there was a "deterioration of the species" as a result of years of hunting, poaching, and famine.[69] An absence of older bulls causes a lack of social order among herds, which may lead to increased fighting among young bulls and an extended rutting season. Additionally, as part of the adaptive genetic selection process, cows given the choice tend to choose mature bulls with larger racks, and without their presence conception rates are reduced and delayed. As a result, if calves are born later in the spring, their feeding season is not long enough to gain sufficient weight for the winter, perpetuating the cycle of winter famine.[70]

In contrast to Leek's images of suffering animals, after the feeding program began, elk were portrayed as welcome visitors in the feedlot, often including the photographer himself and his family surrounded by the animals, who were sometimes shown eating from their hands. In several of these seemingly harmonious scenes of human and animal, the photographer appears with his camera, documenting how close his subjects were. In a self-portrait, Leek is completely surrounded by elk, a large grin on his face, and there is a blur in the foreground of the image: an elk ear perhaps as an animal gets too close to the camera. In another picture, Leek's son Holiday is shown setting up a tripod and camera while swarmed by curious elk, many of which are looking directly at the "father of the elk," who is taking the photograph (fig. 4.3). Of Leek's elk photographs Gregg Mitman writes, "This domestication of elk on screen paralleled the narrative conventions of nature writers like Ernest Thompson Seton and wildlife displays in the national parks. In each case, wildlife was made familiar."[71] The seeming domestication of wild elk contributed to the public's desire to save these animals; by including humans in these pictures to make them appear less wild and more docile, elk appear as "familiar, semi-domesticated pets."[72]

FIG. 4.3. Stephen Nelson Leek, *Holly [Holiday] Leek Photographing Elk on the Leek Ranch, Jackson Hole, WY*, no date. Photographic print, original version glass plate negative, 16.5 × 21.6 cm. S. N. Leek Papers, American Heritage Center, University of Wyoming.

NEW PROBLEMS FACING ELK AND HUMANS

Throughout his career, Leek's work portrayed game animals as subjects of the camera and the hunter, as well as documenting the ecological crises of the elk by bearing witness to winter starvation and poaching. These were animals understood in relation to specific standards "appropriate" to human-animal relations: as subject to hunting by the white sportsman who respected fair chase, as a ward to protect from those who did not follow game laws and requiring intervention from human-caused habitat destruction. But through the anthropocentrism of these sources, the animal experience of the colonial environmental transformation can be pieced together. Leek's documentary photography differed from other wildlife photographers of the time in that he did not seek to erase the presence of humans or the effects of modernity encroaching upon Wyoming. Instead, his photographs and their wide distribution reveal the issues involved in the elk problem, and how supplementally feeding the elk as a solution began a practice that has continued for over a century—which hardly seems like a solution at all.

Since its establishment, the intensely managed Jackson Hole herd of thousands have learned how to navigate highways, roads, and subdivisions as they make their way to the safety of the National Elk Refuge. The sustainable relationship based on practices of subsistence hunting that once existed between Indigenous peoples and animals throughout the region has been replaced with a highly controlled relationship in which elk are artificially fed in the winter and the population managed through licensed hunting. The reliance Euro-American settlers once had on the elk as a local attraction over a century ago has continued in recent years, as evidenced by the iconic elk antler arches that decorate the town square, the establishment of the Museum of Wildlife Art (1987), and annual sleigh rides in the refuge during which tourists can take close-up photographs of elk just as Leek did a century ago. The original circulation of Leek's elk images has expanded to include an engagement in wildlife trade as an important way to keep the refuge running: partial funding of the refuge by annually auctioning naturally shed antlers collected by Boy Scouts.

The success of the institutional feeding program once vaunted by Leek and naturalists has now created problems for natural resource managers in Wyoming. Feeding wildlife for over a century has decreased their migration routes and resulted in overcrowding, increasing the risk of disease, including brucellosis, which has been present in bison and elk of the greater Yellowstone area since 1917, and chronic wasting disease, which was detected in the Jackson elk herd in 2020.[73] What was once a nationally celebrated solution to the elk problem became a problematic practice, and while a new solution has yet to be agreed upon, a structured program for reducing supplemental feeding began in the 2021–22 season.[74] Moreover, elk are still a threat to the livelihood of ranchers: elk continue to compete with livestock for forage, destroy fences, and steal from haystacks. Elk also attract wolves (who have returned), increasing the chance that they will prey upon cattle, while their close contact increases the likelihood of transmitting disease to cattle.[75] Evidently, anthropogenic environmental change has continued to impact the elk herds in western Wyoming and the National Elk Refuge, and the initial solution to the "elk problem" is not a permanent solution. At the same time, the concentration of elk in one area has continued to benefit the tourist economy by attracting visitors to the region as well as bringing in revenue from hunters and state game agencies who purchase licenses and conduct guided hunts. Just like a century ago, the Jackson Hole elk herd is central to the feedback loop of tourism, hunting, and conservation that is vital to the local cultural identity and economy.

NOTES

Acknowledgment: This article was written as part of the VICI project Moving Animals (VI.C.181.010), funded by the Netherlands Organization for Scientific Research (NWO).

1. S. N. Leek, *Zoological Society Bulletin* 17, no. 3 (May 1914): 1118.

2. Alan C. Braddock, "Poaching Pictures: Yellowstone, Buffalo, and the Art of Wildlife Conservation," *American Art* 23, no. 3 (Fall 2009): 40.

3. A. Radyclyffe Dugmore, "A Revolution in Nature Pictures," *World's Work* 1 (November 1900): 36.

4. William Cronon, "Landscapes of Abundance and Scarcity," in *The Oxford History of the American West*, ed. Clyde A. Milner, Carol A. O'Connor, and Martha A. Sandweiss (New York: Oxford University Press, 1994), 605.

5. For more on Indigenous histories of Jackson Hole, see Adam R. Hodge, *Ecology and Ethnogenesis: An Environmental History of the Wind River Shoshones, 1000–1868* (Lincoln: University of Nebraska Press, 2019); Gary A. Wright, *People of the High Country: Jackson Hole before the Settlers* (New York: Peter Lang, 1984).

6. Olaus Johan Murie, *The Elk of North America* (Harrisburg, PA: Stackpole, 1951).

7. Christina M. Cromley, "Historic Elk Migrations around Jackson Hole, Wyoming," in Tim W. Clark et al., *Developing Sustainable Management Policy for the National Elk Refuge, Wyoming*, Yale School of Forestry and Environmental Studies, Bulletin Series 104 (2000): 66–100.

8. Alfred W. Crosby, *Ecological Imperialism* (Cambridge: Cambridge University Press, 2015), 194.

9. Cromley, "Historic Elk Migrations around Jackson Hole, Wyoming."

10. John Daugherty, *A Place Called Jackson Hole: The Historic Resource Study and Grand Teton National Park* (Moose, WY: Grand Teton National Park, 1999), 150.

11. Under the Homestead Act of 1862, ranchers tried to raise cattle on their 160-acre claim, with some adding an additional 160 acres when the Desert Land Act of 1891 was amended.

12. *An Act Establishing Yellowstone National Park* (1872).

13. Paul Schullery, "Yellowstone's Ecological Holocaust," *Montana: The Magazine of Western History* 47, no. 3 (1997): 16–33.

14. Jeffrey C. Mosley and John G. Mundinger, "History and Status of Wild Ungulate Populations on the Northern Yellowstone Range," *Rangelands* 40, no. 6 (December 2018): 189–201.

15. Dillon Wallace, "Saddle and Camp in the Rockies," *Outing; Sport, Adventure, Travel Fiction* 58, no. 2 (May 1911): 196; Theodore Roosevelt et al., *The Deer Family* (New York: Macmillan, 1902), 131.

16. P. W. Norris, *Annual Report of the Superintendent of the Yellowstone National Park to the Secretary of the Interior for the Year 1880* (Washington, DC: Government Printing Office, 1881), 39.

17. Collecting money for wolves was especially lucrative, since you could receive multiple bounties on one animal from organizations such as local stockmen's associations, private ranches, the county, and the state. Vernon Bailey, *Forest Service Bulletin 72: Wolves in Relation to Stock, Game, and the National Forest Reserves* (Washington, DC: US Dept. of Agriculture, 1907): 10; Peter M Zmyj, "'A Fight to the Finish': The Extermination of the Gray Wolf in Wyoming, 1890–1930," *Montana: The Magazine of Western History* 46, no. 1 (Spring 1996): 14–25.

18. Cronon, "Landscapes of Abundance and Scarcity," 612.

19. Today elk can be found in these regions, but these are a different subspecies, the Rocky Mountain Elk, and were transferred from Yellowstone National Park at the beginning of the twentieth century as part of "redistribution" programs to replenish lost populations. See Michael A. Amundson, "'The Most Interesting Objects That Have Ever Arrived': Imperialist Nostalgia, State Politics, Hybrid Nature, and the Fall and Rise of Arizona's Elk, 1866–1914," *Journal of Arizona History* 61, no. 2 (Summer 2020): 255–94.

20. Leek's photographic collection is held in the S. N. Leek Papers at the University of Wyoming's American Heritage Center. Most photographs in the collection are undated.

21. David Stradling, ed., *Conservation in the Progressive Era: Classic Texts* (Seattle: University of Washington Press, 2004), 6.

22. For more on camera hunting, see Finis Dunaway, "Hunting with the Camera: Nature Photography, Manliness, and Modern Memory, 1890–1930,"

Journal of American Studies 34, no. 2 (2000): 207–30; Gregg Mitman, "Camera Hunting," *Reel Nature: America's Romance with Wildlife on Film* (Seattle: University of Washington Press, 1999), 5–25.

23. Theodore Roosevelt, "Introduction," in *Camera Shots at Big Game,* by A. G. Wallihan (New York: Doubleday, Page, 1901), 5, 11.

24. George Eastman Kodak Company advertisements, ca. 1905–7.

25. Robert Sterling Yard, *The National Parks Portfolio* (Washington, DC: Government Printing Office, 1917), 35.

26. This saying is attributed to Chief Seattle (1786–1866) of the Suquamish and Duwamish in the form "Take only memories, leave only footprints" and has come to stand for an ethos encouraged by organizations that promote responsible recreational uses of outdoor spaces.

27. Matthew Brower, "'Take Only Photographs': Animal Photography's Construction of Nature Love," *Invisible Culture: An Electric Journal for Visual Culture*, no. 9 (2005), https://ivc.lib.rochester.edu/take-only-photographs-animal-photographys-construction-of-nature-love/.

28. Matthew Brower, *Developing Animals: Wildlife and Early American Photography* (Minneapolis: University of Minnesota Press, 2011).

29. For instance, *Camera Shots at Big Game* includes several photographs of mountain lions treed by dogs, with accompanying text detailing how several of the lions were killed after their pictures were taken. Allen Grant Wallihan and Mary Augusta Wallihan, *Camera Shots at Big Game* (New York: Doubleday, Page, 1901).

30. This game law continued in 1915. Frank L. Houx, *Game, Bird, and Fish Laws of Wyoming, 1915–1916* (Cheyenne: Wyoming Labor Journal, 1915), sec. 22.

31. Matthew Brower, "George Shiras and the Circulation of Wildlife Photography," *History of Photography* 32, no. 2 (2008): 170.

32. S. N. Leek, *The Elk of Jackson's Hole, Wyoming: Their History, Home and Habits* (Jackson, WY: S. N. Leek, 1914), available at Jackson Hole Historical Society and Museum, 2002.0133.053a.

33. Karl Jacoby, *Crimes against Nature: Squatters, Poachers, Thieves, and the Hidden History of American Conservation* (Berkeley: University of California Press, 2001), 139.

34. *Fort Bridger Treaty Council of 1868*, Article IV.

35. John Clayton, "Who Gets to Hunt Wyoming's Elk? Tribal Hunting Rights, U.S. Law and the Bannock 'War' of 1895," *Wyoming State Historical Society*, 2020.

36. US Bureau of Indian Affairs, *Annual Report of the Commissioner of Indian Affairs, 1895* (Washington, DC: Government Printing Office, 1895), 63–68.

37. *Ward v. Race Horse* declared that the law regulating Bannock hunting rights

"does not give them the right to exercise this privilege within the limits of that state in violation of its laws." Ward v. Race Horse, 163 U.S. 504 (Supreme Court of the United States, 1896), 504.

38. "The Silly Season War," *Forest and Stream* 45, no. 5 (August 3, 1895): 89.

39. "Indian Hunting Rights," *Forest and Stream* 46, no. 23 (June 6, 1896): 449–51; S. N. Leek, letter to the editor, *Forest and Stream* 46, no. 23 (June 6, 1896): 455.

40. S. N. Leek, letter to the editor, *Recreation* 5, no. 6 (1896): xvii.

41. See Jan Dizard, *Going Wild* (Amherst: University of Massachusetts Press, 1994); Dizard, *Mortal Stakes: Hunters and Hunting in Contemporary America* (Amherst: University of Massachusetts Press, 2003).

42. Thorstein Veblen, *The Theory of the Leisure Class* (New York: Macmillan,1899), 258.

43. "Herds of Elk Slaughtered: Animals Caught in Deep Snow and Butchered," *Los Angeles Times*, December 16, 1906.

44. Elk teeth have traditionally been used by many Indigenous tribes, including the Sioux, Crow, Assiniboine, and Cheyenne, as a form of embellishment on garments, jewelry, and objects. For some cultures elk eyeteeth symbolize longevity, since they do not decay like the rest of the animal's teeth and body, and the Bannock and Shoshone used them as a form of currency. Henry Balfour, "Note on the Use of 'Elk' Teeth for Money in North America," *Journal of the Anthropological Institute of Great Britain and Ireland* 19 (1890): 54; Josephine Paterek, *Encyclopedia of American Indian Costume* (New York: W. W. Norton, 1994).

45. Elizabeth Wied Hayden, "Driving Out the Tuskers," *Teton: The Magazine of Jackson Hole* (Winter–Spring 1971): 22–36.

46. Karl Jacoby discusses the politics of elk poaching in the Yellowstone area in *Crimes against Nature*, 133–40. Robert B. Betts, *Along the Ramparts of the Tetons: The Saga of Jackson Hole, Wyoming* (Boulder: Colorado Associated University Press, 1978), 184; Murie, *The Elk of North America*, 286.

47. Esther B. Allan, "History of Teton National Forest," unpublished manuscript, USDA Forest Service, 1973, 137.

48. The same photographs were recycled and attributed to Frank Jay Haynes to illustrate an editorial in *Forest and Stream* as evidence of illegal poaching that took place in Yellowstone in 1894. See Alan C. Braddock, "Poaching Pictures: Yellowstone, Buffalo, and the Art of Wildlife Conservation," *American Art* 23, no. 3 (Fall 2009): 36–59.

49. Specifically, Timothy O'Sullivan, *The Harvest of Death: Union Dead on the Battlefield at Gettysburg, Pennsylvania*, July 5–6, 1863.

50. Thomas L. Altherr, "The American Hunter-Naturalist and the Development of the Code of Sportsmanship," *Journal of Sport History* 5, no. 1 (1978): 7–22.

51. "Slaughtering Elk for Their Teeth," *Forest and Stream*, January 1902, 30; Roosevelt, "State of the Union Address," December 2, 1902.

52. S. N. Leek, *Recreation* (1900), 102–3.

53. S. N. Leek, *Recreation* (1900), 192.

54. S. N. Leek, "The Problem of the Elk," *Collier's Weekly* 46, no. 2 (February 1911): 22.

55. L. M. Brett, *Report of the Acting Superintendent of the Yellowstone National Park to the Secretary of the Interior* (Washington, DC: Government Printing Office, 1911), 9.

56. Wallace, "Saddle and Camp in the Rockies," 195.

57. W. A. Bartlett, "Forest and Stream," *Forest and Stream* 72 (1909): 458.

58. "The Elk Problem," *Forest and Stream* 77 (1911): 437.

59. Wallace, "Saddle and Camp in the Rockies," 188.

60. "Wyoming Elk Starving in the Midst of Plenty," *Semi-Weekly Boomerang*, March 13, 1911.

61. Frederic Irland, "The Wyoming Game Stronghold," *Scribner's Magazine* 34, no. 3 (September 1903): 262.

62. William A. Frassanito, *Gettysburg: A Journey in Time* (New York: Charles Scribner's Sons, 1975), 186–92.

63. Wallace, "Saddle and Camp in the Rockies," 188.

64. Caspar Whitney, ed., "The Sportsman's Viewpoint," *Collier's Weekly* 46, no. 2 (February 1911): 24. For the story of how elk from Yellowstone were brought to Arizona, see Robert N. Looney, with commentary by Neil Carmony, "How Arizona Was Restocked with Elk," *Arizona Wildlife Views* 46 (January–February 2003): 24–28.

65. Irland, "The Wyoming Game Stronghold," 262.

66. Irland, "The Wyoming Game Stronghold," 262.

67. S. N. Leek, "A Significant Letter about Elks: Letter to Governor Brooks," *Wyoming Industrial Journal*, April 1, 1909.

68. Hornaday joined Leek's appeal to Congress to pass a bill to support the elk in 1911. "5,000 Elk Starving; Congress Aid Sought," *New York Times*, February 12, 1911.

69. William Temple Hornaday, *Our Vanishing Wildlife: Its Extermination and Preservation* (New York: Charles Scribner's Sons, 1913), 338.

70. James H. Noyes et al., "Effects of Bull Age on Conception Dates and Pregnancy Rates of Cow Elk," *Journal of Wildlife Management* 60, no. 3 (1996): 508–17.

71. Mitman, “Camera Hunting,” 94.

72. Mitman, “Camera Hunting,” 94.

73. Nathan L. Galloway et al., “Supporting Adaptive Management with Ecological Forecasting: Chronic Wasting Disease in the Jackson Elk Herd,” *Ecosphere* 12, no. 10 (2021); Wyoming Game and Fish Department, *Jackson Bison Herd (B101) Brucellosis Management Action Plan* (Cheyenne: Wyoming Game and Fish Department, 2008).

74. US Fish and Wildlife Service, *Bison and Elk Management Step-Down Plan: National Elk Refuge, Grand Teton National Park,* (Lakewood, CO: Department of the Interior, 2019).

75. Mosley and Mundinger, “History and Status of Wild Ungulate Populations.”

JESSICA WANG

five Animals, Infrastructure, and Empire Insects and Birds as Biological Control Agents in Early Twentieth-Century Hawai'i

In 1898, the Territory of Hawaii became part of a US overseas empire, following the annexation of the Republic of Hawaii, itself the creation of a coup against the native Hawaiian monarchy five years earlier organized by the kingdom's sugar elites and backed by US military power. In an island economy centered on agriculture, US rule meant not just the establishment of political authority but extensive initiatives to manage ecologies and landscapes from the basic biological standpoint of regulating animals' presence. Maintenance of forage for horses and livestock, the defense of the sugar industry from invasive insect species, the protection of cattle from insect pests, the safeguarding of forest ecosystems against goats brought in to augment subsistence and trade or from invasive insects that threatened native trees, and myriad other concerns demanded constant attention from Hawai'i's territorial government. Such diverse activities suggest the extent to which colonialism depended not only on the assertion of claims of sovereignty over territory and Indigenous peoples but also active efforts to manage and oversee unruly and ever-shifting biological relationships.

The ecological dimensions of animal-based governance speak to and expand upon a number of themes addressed in this volume, particularly the intersections between animals, the state, and the evolving economic relationships and infrastructures of the nineteenth and early twentieth centuries. Animals' significance as commodities made them into objects of human concern, and just as dying cattle in Montana forced ranchers to take responsibility for the subsistence of their animal charges, so too did horn fly infestations lead ranchers and agricultural officials to seek means of preserving the value of animal bodies. Managed elk populations in Wyoming also reflected the combined motives of economic interest and sympathy for charismatic animals' suffering. Animals constituted infrastructure as well, as Jennifer Marks's account of Chicago adeptly demonstrates. Whereas horses' infrastructural role grew out of their status as commodified labor, however, hymenoptera and birds introduced at public expense in the Territory of Hawaii underscored the significance of the state and its effort to establish an ecological infrastructure capable of protecting cattle and forests. Indeed, by attempting to engineer the equilibrium among animal populations, the state treated the environment itself as a key infrastructure that conditioned US colonial order.

With this idea of animal-based infrastructure in mind, this chapter focuses on two episodes from the early twentieth century. Horn fly control efforts in the 1910s and discussions about bird introductions in the 1920s draw attention to how overlapping, globalized networks of multiple species shaped attempted introductions of insects and birds as biological control agents in early twentieth-century Hawai'i. First, the nexus of cattle and horn flies led agricultural officials to seek out insect predators, followed by avian alternatives, from far-flung parts of the world. Discussions of proposed bird introductions also intersected with a second problem: the threat of invasive insects to forest ecosystems. Despite birds' potential threat to the so-called beneficial insects that the territorial government had brought in for purposes of insect pest control, in the 1920s experts began to consider intentional bird introductions from Australia, Asia, and elsewhere to defend the koa forests of Maui.

The human activities that brought cattle, horn flies, other invasive insect species, and insect and avian countermeasures to the Hawaiian Islands constituted a high-stakes game of ecological whack-a-mole—an animal metaphor chosen quite deliberately here to emphasize how every new species introduction generated a chain of consequences that perpetuated the colonial state's system of oversight and intervention. Agricultural experts hoped that deliberate introductions of specific animal species could protect and maintain food systems, the trade in commodified animals, and the general climate and environment of the US-claimed Territory of Hawaii. Pretensions of human management, however, failed to acknowledge more fundamental biological realities, in which humans participated more or less coequally alongside other species in the networks of animal mobility and exchange that undergirded global capitalism and its associated ecologies.

IN SEARCH OF BIOLOGICAL CONTROL: HORN FLY PARASITE INTRODUCTIONS, 1913

In July 1913, W. G. Ogg, manager of the Hawaiian Agricultural Company, pleaded with his operation's parent corporation for advice about chemical applications that might help in dealing with horn flies and their depredations on local cattle. He wrote plaintively, "We have never seen this noxious insect in such numbers and many of our animals have large raw spots on their backs and sides which are a disgrace to be seen."[1] The problem was not merely aesthetic or a matter of livestock animals' comfort. In dairy cows and beef cattle, severe horn fly infestations led to weight loss and reduced milk production, which posed a serious economic threat.[2]

Introduced species and their ecological consequences were an integral part of the natural and human history of Hawai'i, in which the era of US imperial rule constituted only the latest phase of the islands' ecological development. Life in the Hawaiian Islands evolved as a process of biological colonization from the beginning. Barren volcanic formations gained new life forms when seeds and insects hitched rides on migratory birds or flotsam on the ocean, winged species established themselves on new and rocky shores, and semi-aquatic animals found breeding sites or resting places suited to their requirements. Starting around the year 1300, human arrivals from Polynesia further transformed the islands' plant and animal populations when they brought taro, sweet potatoes, breadfruit, and other crops, along with pigs, chickens, and dogs. Hundreds of years later, when humans from Europe, the United States, and elsewhere arrived in Hawai'i in the late eighteenth and early nineteenth centuries, they admired the sophistication of Native Hawaiian agriculture and aquaculture, as well as the islands' ideal location as a port of call and their seeming potential for "improvement" through colonial intervention. With Hawai'i's nineteenth-century integration into the global economy, the islands entered a new stage of ecological transformation through the settlement of foreign missionaries, the islands' importance as a stopping point for whaling ships, and agricultural enterprises that included the boom and bust of the sandalwood trade and the early rise of the sugar industry.[3]

The introduction of novel animal species, whether by accident or design, was an integral part of Hawai'i's global status. Discomfited by lands they perceived as animal-poor, as well as a lack of birdsong in the islands' aural environment, European and American settlers brought in livestock animals, and in the second half of the nineteenth century, they also pushed for new bird species, whether as sources of game or for purposes of beautification.[4] Species intentionally admitted frequently wreaked unanticipated environmental havoc. The nineteenth-century transportation revolution, in which steam power rapidly accelerated the pace of travel, also vastly enhanced insects' mobility. Unwanted insects now found it possible to survive long ocean voyages and threaten agricultural endeavors when they colonized new places. This newly heightened threat from invasive species, combined with an era of growing expectations of governmental power, prompted determined efforts to oversee and control species introductions through inspections, quarantines, and other regulatory measures. California and other Pacific Coast states, along with the independent Kingdom of Hawai'i governed by the Native Hawaiian monarchy, all implemented plant inspection and quarantine regimes in the 1880s and 1890s, and the federal government eventually followed suit with the Plant Quarantine Act of 1912.[5] With the implementation of US colonial rule in Hawai'i, the territorial government's Board of Commissioners of Agriculture and Forestry (BCAF) quickly established its determination to avoid or contain damage to agriculture and to the islands' forests by preventing the introduction of harmful species or, where necessary, implementing biological countermeasures.

The BCAF's and ranchers' efforts to deal with the horn fly problem exemplified this early twentieth-century aspiration to control invasive species. By the time of Ogg's appeal, horn fly had been a recognized problem in the Hawaiian ranching and dairy sector for fifteen years. Although more recent sources have guessed that *Haematobia irritans* could have arrived along with the first cattle on the islands in 1793, late nineteenth-century observers carefully documented the horn fly's recent appearance. The latter's reckoning seems more persuasive, in light of the transportation revolution of the second half of the nineteenth century and its role in spreading invasive species. In 1898, entomologist Albert

Koebele reported to the minister of the interior of the Republic of Hawaii his firm conviction of the fly's recent arrival. The US Department of Agriculture had identified it as a new introduction from Europe to the Philadelphia area in the mid-1880s, and from there, the horn fly spread rapidly throughout North America. In Hawai'i, Joseph Paul Mendonça, co-owner of the Kaneohe ranch on O'ahu and a civic leader who had participated in the overthrow of the Hawaiian monarchy, which eventually led to the US annexation of the islands, first reported the horn fly's appearance on his ranch in February 1898. Koebele stated categorically, "The fly has without doubt come over with cattle from the west coast." He estimated that it had landed two or three months before attracting Mendonça's notice.[6]

For American agricultural experts in the late nineteenth century, biological control measures, particularly the introduction of parasites to control the numbers of unwanted insect species, provided the dominant means of dealing with invasive pests. Charles V. Riley, head of the Bureau of Entomology at the US Department of Agriculture (USDA), first became interested in the idea of biological control in the 1870s, and he had the opportunity to direct the search for insect "enemies" to deal with an outbreak of cottony cushion scale in California's citrus groves in the late 1880s. Koebele was working for the USDA at the time, and the agency sent him to Australia, where he quickly identified vedalia (*Novius cardinalis*), a ladybird species, as a promising candidate for dealing with California's citrus pest. The introduction of vedalia achieved extraordinary results, and the cottony cushion scale virtually disappeared from California orchards within a year. Biological control efforts have rarely scored such dramatic success in the decades since, but the fortuitous choice of vedalia established the strategy as the leading option for insect pest control throughout the early twentieth century. This new era of biological control also generated professional opportunities for entomologists, who became key personnel in US agricultural institutions in both national and colonial settings.[7]

Agricultural officials in Hawai'i, often in collaboration with entomologists on the staff of the experiment station of the Hawaiian Sugar Planters Association (HSPA), regularly sponsored expeditions to seek "beneficial insects" to control various pests, including the horn fly. In response to the vedalia triumph, the provisional government of the Republic of Hawaii quickly poached Koebele from the USDA in 1893, and he undertook frequent entomological explorations for either the official agricultural bureaucracy or the HSPA experiment station for the next two decades. Hence Koebele was on hand to describe the horn fly's life cycle upon its first detection in Hawai'i and to propose potential countermeasures. In subsequent years, he collected dung beetles and other species in the US Southwest, as well as other parasites in Europe. Unfortunately, the attempted introductions either failed to take or had little effect on the horn fly's burgeoning numbers.[8]

A new and more ambitious parasite release effort took place in 1913–14, and it illustrates the globalized networks and social relationships that undergirded biological control efforts and their human-mediated species introductions. In 1912, the BCAF hired the Italian entomologist Filippo Silvestri to undertake an expedition to West Africa in search of parasites to combat a severe outbreak of Mediterranean fruit flies, which were destroying a vast range of garden fruits in the Hawaiian Islands. Silvestri began his journey in July, and he arrived the following May in Honolulu with a number of candidate species acquired along his route from West Africa to South Africa, Australia, and finally, Hawai'i.[9] His precious living cargo also included a gift from Charles Mally of the Cape Colony's Entomological Office. Thanks to Mally, Silvestri landed in Honolulu with samples of *Muscidifurax vorax* (today classified as *Muscidifurax raptor*), a tiny Chalcid wasp known to parasitize the pupae of house flies and horn flies.[10]

Under the supervision of entomologist David T. Fullaway, the new parasite bred readily on horn fly pupae in the BCAF's insectary, which was reliably producing hundreds of parasites a week by early July 1913. The pace roughly doubled in August, and at the height of the project in November, the BCAF was breeding as many as 1,900 *Muscidifurax* in a single day. The ease of the parasites' reproduction raised Fullaway's hopes that if *Muscidifurax* established itself in the wild, it would "be a very efficient check on the horn fly." Its successful introduction, however, required developing appropriate release methods on Hawaiian ranches, and even then, there were no guarantees that *Muscidifurax* would ultimately make a major dent in the horn fly population.[11]

As much as insects might appear to proliferate in new and friendly habitats with unobstructed ease, in reality, populations did not necessarily establish themselves quickly or easily. The horn fly had needed more than a century of cattle importations and the aid of steamships before it managed to make itself at home on the islands.[12] Even intentional efforts at establishing new insect species, namely the parasites that agricultural experts relied upon as biological control agents, required close attention and aid, which made humans a crucial adjunct to the parasites' life cycle. Successful introductions generally necessitated optimal conditions, repeated attempts, large numbers of individual organisms, and a certain amount of ecological good fortune. In September 1913, BCAF president Walter M. Giffard vividly described to his fellow commissioners the challenge for *Muscidifurax vorax*: "To be successful the natural conditions similar to those which the parasite has been accustomed to in its native habitat must be as near perfect as possible and even then the percentage of efficiency of any one parasite may be quite small."[13]

Muscidifurax's prospects for success thus depended upon human methods and networks, especially among agricultural experts and ranchers, to create receptive local environments. Ideally, as Giffard explained to one rancher on Maui, the BCAF preferred to rely on its own entomologists to choose sites and liberate the parasites.[14] With only a small handful of entomologists on staff, however, the board could not dispatch an expert to the field in each and every case. The BCAF therefore relied on a form of citizen science, in which it sent written instructions along with parasite shipments and imposed reporting requirements to find out how the released parasites fared.

Correspondence between ranchers and the Board of Commissioners of Agriculture and Forestry testifies to recipients' eagerness to acquire *Muscidifurax* and their willingness to send in their findings. The release program concentrated on large ranches because, as Fullaway later explained, "these offered exceptional opportunities for [the parasite's] establishment and also because it was felt that these ranches had first call on an enemy of the horn fly, in view of their large interests."[15] In the small social and political world of the Territory of Hawaii, agricultural officials also likely had confidence in the large ranchers' reliability, as a result of regular contact and mutual interests. *Muscidifurax* provided an occasion for strengthening these relationships.

As the *Muscidifurax* breeding program ramped up its pace in August 1913, the board began to send out vials of horn fly parasites with a standard letter that asked for three pieces of information: the location of the parasite liberation site, a description of "conditions as to dung, horn fly maggots and pupae," and whether cattle were being kept away from the dung piles as instructed.[16] I suspect that in most of these cases, ranchers and their representatives had already received more detailed instructions in person or via other correspondence, which had directed them not to release the *Muscidifurax* maggots in full sun, to ensure the proper level of moisture for the manure, and to keep cattle and other animals from trampling in it.[17] For example, in early September, Giffard explained to Palmer P. Woods of Kohala, on the "Big Island" of Hawai'i, the method and its rationale. Release sites needed to be prepared several days in advance, to ensure the availability of horn flies as pupae, which the female *Muscidifurax* needed to lay their eggs and to provide food for the larvae. "The *latter* stage is absolutely essential," Giffard sternly emphasized. "The labor and cost of rearing these parasites is considerable and it will not do to waste a colony on any Ranch if conditions for the time being are not satisfactory." The initial release, which would happen in the presence of Herbert T. Osborn, an entomologist temporarily assigned from the Hawaiian Sugar Planters' Association to assist the BCAF, also constituted a learning experience in preparation for future liberations. Thus Giffard advised, "We suggest that you familiarize yourself with the conditions suitable to your particular locality and under which the parasites may be liberated whilst Mr. Osborne [*sic*] is with you and we can then later on send you further colonies to Mahukona in care of the Purser."[18] In short, once properly supervised and trained by Osborn, Palmer could be trusted to oversee future parasite releases on his own.

For their part, recipients took to their BCAF-mandated fieldwork with enthusiasm. For example, from the settlement at Kalaupapa, Moloka'i, which served as a home and place of forced isolation for nearly nine hundred sufferers of Hansen's disease, facility superintendent J. D. McVeigh sent optimistic news about the parasites: "I am confident that the little fellows are doing good work." He forwarded to the board a detailed report from Emil van Lil, a lay brother who maintained the farm and dairy at Kalaupapa. Van Lil described with precision the preparation of a manure box and liberation of the parasites in accordance with the board's instructions, as well as subsequent results as they unfolded over the next four weeks. "I believe that there is already a decrease of the flies on the stock," he reported happily. He then added a detail that suggested his care in following the board's advice to the letter: "I must not omit to mention that the box was placed under the trees of my yard and was well sheltered from sun and wind." This methodical approach soon yielded highly satisfying results, although van Lil worried that bad weather might undo all of his efforts. "About ten days after planting the bugs," he observed, "the little fellows began to leave the box by the thousands. Up to the present day they are still multiplying but in lesser numbers than they did a week or two ago. My only fear now is that the stormy weather we are having will injure if not perhaps wipe out the colonies."[19] A month and a half later, van Lil feared that the storms had killed off the parasites. He requested a fresh batch, and to demonstrate his worthiness, he also relayed confident plans for sheltered manure boxes that would allow *Muscidifurax* to prosper this time around. The board's superintendent of entomology, Edward M. Ehrhorn, sent a reassuring response that the parasites could generally take care of themselves in rainy weather but nevertheless arranged to dispatch a new lot. Van Lil immediately released the new shipment under good weather conditions: "A breeze from the trades with a clear sunny sky."[20]

Such correspondence speaks to the level of collaboration between agricultural experts and lay constituencies required for biological control efforts in early twentieth-century Hawai'i. In a colonial setting of close ties among local elites, ranching and dairy interests mixed easily with agricultural officials. At times, mistakes happened. More than one writer reported regretfully that livestock interfered with manure piles, either because animals overcame barriers or ranch managers missed key components of the BCAF's directives.[21] Occasionally, signals got crossed, and parasites arrived before ranchers and dairymen had time to prepare the necessary release sites.[22] Sometimes the board impatiently waited for reports and resorted to nagging reminders for follow-up so that it could chart the progress of the horn fly parasite release program.[23] Such snafus aside, however, limited personnel, along with the close-knit social networks that tied the Territory of Hawaii's economic and political elites, meant that the BCAF had every motive to work productively with its constituents. Ranchers and dairy managers also desperately wanted a solution to their horn fly woes. As one correspondent from Puuwaawaa (Pu'uwa'awa'a) Ranch wrote following his release of the parasites, "We certainly hope that it proves a success as horn flies are very bad here."[24]

Not surprisingly, people who earned their livings off the land and livestock management also frequently proved adept observers of the parasites under their charge. From Moloka'i, for example, a Mrs. Knott reported to Fullaway about the two different sites she prepared and how the maggots fared: "One of the heaps of manure I protected from the sun, & the other I left unsheltered. The one under cover did not do so well as the other." In essence, Mrs. Knott took it upon herself to conduct her own experiment and to identify shortcomings in the board's procedures. She also noted the failure of an initial release to take and asked whether "the ants interfere with the parisite [*sic*]?"—a highly perceptive remark, since entomologists knew that a particular ant species in Hawai'i preyed upon the pupae of *Haematobia irritans*.[25] Other reports also demonstrated laypersons' acuity. At the Kukaiau Ranch on the Big Island, for example, a keen-eyed ranch manager, Donald S. Macalister, detected other parasite species at work in the manure, namely, the descendants of Albert Koebele's introductions years earlier. "If I am not mistaken," Macalister wrote, "it is the same kind of parasite as was introduced into this country some six or seven years ago to fight the Horn Fly."[26]

Indeed, some ranchers had as much entomological expertise as agricultural officials in an era of blurred boundaries between professional and amateur naturalists. Ranch manager George C. Munro wrote in from Lanai to document the careful conditions under which he had released the board's parasites.[27] Munro came from New Zealand, and his ample experience as a naturalist included early travels as far as Midway to collect bird specimens in the 1890s and membership in the New Zealand Institute (now the Royal Society Te Apārangi). His biographical entry in *Men of Hawaii* described him as "associated with natural history investigations and museums since early youth, discovering new species in bird, insect, and plant life," and his explorations and affiliations tellingly placed Munro within a set of trans-Pacific connections that tied Hawai'i to Australasia.[28] This background provided Munro the authority and cachet to communicate with board personnel as a peer. Notably, the British-born Giffard, who had worked his way up the economic and social ladder in W. G. Irwin & Company and the business world of Hawaiian sugar, himself possessed a not dissimilar life story and built a powerful entomological reputation without formal training. An avid collector, he helped found the Hawaiian Entomological Society, and he described numerous insect species not previously identified as he moved from the sugar industry to a career with the BCAF.[29] The shared entomological enthusiasms and sense of fellowship between the two men meant Munro could enjoy unquestioned status as a reliable recipient of *Muscidifurax*, someone whom the board could count on to ensure the parasites' proper release, as well as knowledgeable follow-up reporting.

The BCAF wound down this initial *Muscidifurax* breeding program at the end of 1913.[30] In subsequent years, the board continued to try to counter the horn fly by releasing more *Muscidifurax*, small numbers of a second chalcid species, and two other parasite species from the Philippines, another key Pacific outpost of the post-1898 American empire.[31] A complex assemblage of people and parasites brought the latter to Hawai'i, but as with Koebele's earlier parasite introductions, these new additions to the islands' insect population failed to make much of a dent in the horn fly problem. In the immediate post–World War I period, strategy shifted toward dung reduction, through new entomological expeditions aimed at bringing in species of dung beetles that could work to deny breeding grounds to the horn fly.

Birds offered another possibility. As Fullaway noted in July 1919, in addition to dung beetles, "it is also believed that the predation of insectivorous birds is an aid which must be employed to bring about a reduction of the fly."[32] As an issue of "paramount" importance, cattle mattered enough to the Territory of Hawaii's economy that agricultural experts began to consider bird introductions that they had adamantly rejected in prior decades.[33] Mounting problems with invasive insects in delicate forest ecosystems further boosted birds' prospects as biological control agents. Aesthetic, economic, and biological considerations melded in the debate over bird introductions in the 1920s.

PROPOSED BIRD INTRODUCTIONS: FROM ECONOMIC STANDARDS TO THE THREAT OF "BIOLOGICAL COMPLEXITY"

Proposals to introduce particular species of birds shed light on the varied disciplinary, institutional, and ecological balancing acts involved in attempting to maintain the territory's agricultural interests as well as the multiple species necessary to the basic health of island ecosystems. The possibility of employing birds as biological control agents had long interested bird enthusiasts, and acclimatization societies started promoting birds' usefulness—along with their aesthetic qualities and even their moral worthiness as hardworking insect-consumers—in the mid-nineteenth century. Unbridled enthusiasm waned, however, with some prominent instances in which introduced avians disrupted local ecosystems and threatened native species.[34] From the perspective of BCAF officials in the early twentieth century, controlled bird introductions promised benefits ranging from insect pest control to forest preservation, but risked the proliferation of new invasive species if experts erred in their assessment of newcomers' likely habits and habitats. The zeal of sportsmen for game birds and bird enthusiasts' desire to beautify the islands' landscape and soundscape according to Euro-American standards also suggested the unspoken colonial mentalities at work in early twentieth-century debates about bringing new bird species to the Territory of Hawaii.

Bird introductions came under the purview of the BCAF's Rule II, established in October 1904, which forbade the introduction to the Territory of Hawaii of certain bat and crustacean species, along with "any other animal, bird, reptile or insect injurious, or liable to become injurious to forests, trees, plants or other vegetation of value."[35] As this language indicated, economic considerations determined the standard for whether to bar specific organisms for the sake of plant protection, but, as some BCAF officials pointed out, Rule II did not provide much in the way of specific guidance. In the early 1910s, when the question of bird importations became increasingly pressing, the board deliberated over whether to adopt new, more exacting regulations.[36]

In response to the "constant demand" to introduce songbirds and insect-feeders to Hawai'i, Walter M. Giffard chaired a Committee on Entomology to consider the need for regulatory reform, and he immediately consulted with experts on whether some bird species might be admitted safely.[37] Experts' responses highlighted the unknowns of how introduced birds might behave in a new climate and habitat, as well as the need to protect insect parasites acquired for biological pest control. William A. Bryan, an ornithologist and professor of zoology at the College of Hawaii in Honolulu, was "keen to have bird introduction commenced here without delay," but only based on proper professional evaluation. At the same time, he acknowledged that no good scientific research existed that could predict with certainty how birds from European or North American climates would behave in Hawai'i. Nonetheless, Bryan enthusiastically supported the introduction of so-called beneficial bird species—especially insect-feeders that would not threaten human food production—following judicious experimentation to determine how they would fare in the islands.[38] Naturalist Robert C. L. Perkins, director of the entomology division of the Hawaiian Sugar Planters' Association Experiment Station and an expert in both entomology and ornithology, adopted the more cautious stance that new bird populations should be established only "owing to some extreme necessity," and even then not without "prolonged and extensive experiments under comparatively natural conditions." In particular, he warned that insectivorous species would consume not just undesired pests but also the introduced insect parasite species brought to the islands at such high expense and effort. From Perkins's standpoint, ornithologists too often did not take such concerns into sufficient account until enlightened by entomological experts.[39]

Based on Bryan's and Perkins's input, the Committee on Entomology reaffirmed the status quo in a manner that upheld the importance of expertise while simultaneously recognizing the limits of available resources. The risks associated with bird introductions, the committee reported, meant that decisions could not be left to private individuals. The lack of funds for research and experimentation, however, meant that the board had no means of determining which "birds are likely to be of use for the purposes required and at the same time guarantee such birds against depredations of the greatest economic importance."[40] This position left the BCAF to reject new regulations for lack of means to administer them but also provided a rationale for the board's subsequent conservatism when faced with new requests for bird importations. For example, when Hilo lawyer William S. Wise pushed for more songbirds and insectivorous birds in Hawai'i, Giffard could rely on the expert judgment of "the two best ornithologists available in the islands" to convey regrets that the board could not pursue the experiments necessary to ensure safe introductions of new avian fauna.[41] When Wise and Bryan, along with Walter E. Wall, government surveyor and a leading member of the Hawaiian Fish and Game Association, suggested that a California wren species could help combat unwanted insect populations, Giffard cited prospective parasite introductions as a major reason to keep the wren out.[42] In response to a USDA ornithologist's suggestion that three species of game birds from Mexico could be strong candidates for Hawai'i, the BCAF's superintendent of forestry, Ralph S. Hosmer, also sounded a note of extreme caution. He admonished, "It is the policy of the Board to be very careful about introducing into Hawaii any birds liable in any way to disturb the favorable conditions brought about thro' the establishment of beneficial insects."[43] In the absence of experimental data beyond the means of the BCAF to provide, the welfare of laboriously established hymenoptera and other insect parasites had to take precedence over new birds in the islands.

By the 1920s, however, the board became more receptive to bird introductions as persistent insect problems opened the possibility for avian countermeasures. On the one hand, agricultural officials remained cautious, and by the late 1910s, federal authority added greater specificity to territorial provisions.

As Edward W. Nelson of the USDA's Bureau of Biological Survey informed Edward M. Ehrhorn in 1919, the prohibitions against any bird "likely to prove injurious to agriculture" included no weaver birds or members of the family Fringillidae whatsoever, as well as "no other bird of which there is any doubt."[44] On the other hand, however, damage by insects to agriculture and to the forests that maintained Hawaiian watersheds began to shift the balance of interests in favor of novel birds.

Discussions about the potential introduction of an Australian flycatcher species, particularly David T. Fullaway's position, indicate the recalibrations at work. Following his participation in parasite-breeding efforts, Fullaway accepted a full-time position with the BCAF, and his subsequent collecting ventures included the board's second West African expedition to search for additional Mediterranean fruit fly parasites in 1914–15, and then an expedition to Asia to seek out melon fly parasites in 1915–16. By the 1920s, Fullaway increasingly looked to birds for insect pest control. As part of the transpacific nexus of agricultural exchange that linked California, Hawai'i, and Australia, Fullaway wrote to James Franklin Illingworth, a former entomology professor and ornithological enthusiast at the College of Hawaii who left to take a post with the Bureau of Sugar Experiment Stations in Queensland, Australia, for advice on bird species that might help with insect control in Hawai'i.[45] Illingworth suggested *Rhipidura tricolor* (today reclassified as *Rhipidura leucophrys*), known more commonly as either the willy wagtail or shepherd's companion, as "a most valuable bird for Hawaii."[46] Although zoologist Joseph Grinnell of the University of California's Museum of Vertebrate Zoology warned Fullaway that resorting to introduced birds in order to keep insects in check posed major risks to native species, Fullaway's reading of the scientific literature encouraged him to embrace the willy wagtail as a way to deal with the ongoing horn fly problem.[47]

Frederick A. G. Muir, who as one of the HSPA's leading entomologists had conducted three successive expeditions in Southeast Asia, China, and the South Pacific from 1906 to 1908, which ended with the successful introduction of a tachinid fly parasite to deal with the sugarcane borer, supported Fullaway's initiative, as did the director of the HSPA's Experiment Station, H. P. Agee.[48] Although the islands' agricultural professionals always emphasized that avian importations should serve economic purposes only and not aesthetic objectives, birds' charismatic qualities nonetheless seeped into discussions.

For Europeans and Americans accustomed to aural environments rich with birdsong, the relative paucity of avian species in the Hawaiian Islands disturbed foreign settlers in the mid-nineteenth century, and enthusiasts frequently called for the importation of songbirds as a civic beautification measure.[49] Territorial agriculture officials refused the introduction of birds for purely aesthetic purposes, but when their admission could be justified economically, experts proved just as vulnerable to individual species' particular charms. As Muir tellingly noted, not only did *Rhipidura tricolor* have the appetite to make inroads on the horn fly population, but in addition, "it is a pretty little bird and spreads its tail out fan-like and wags it from side to side." It earned its name "shepherd's companion" in the Australian outback, where "in those lonely places [it] is looked upon with kindly feelings by the shepherd and stock men."[50] Other species with promise to combat insects attracted similar praise, such as the "dainty mincing strut" of Australia's magpie lark, or the "ornamental plumage, pleasing song, and small size" of the Pekin nightingale.[51]

With the assurances of Muir and Agee, the BCAF authorized the introduction of the willy wagtail, and with HSPA entomologist C. E. Pemberton already in Australia on other business, the board requested Pemberton's assistance in arranging for the birds' delivery. Australian officials, however, discouraged the acquisition due to the difficulty of capturing the birds and their likelihood of not surviving the voyage to Hawai'i. Four years later, the board finally succeeded in gaining Australian permission, and the first shipment of *Rhipidura tricolor* left Sydney in 1926.[52] In the 1920s, the BCAF also approved other transpacific bird introductions in order to deal with the horn fly, the cut-worms that plagued garden crops, and other "injurious insects." The new avian arrivals included the straw-necked ibis and magpie lark from Australia, and also the California meadowlark. In addition, the board backed introductions to Maui of two species already established elsewhere in the islands: the Chinese huamei, which colonized O'ahu following its accidental release during the Honolulu fire of 1900, and the Japanese tom-tit, which had made itself at home on Kauai.[53]

Meanwhile, the deterioration of native forest habitats led concerned citizens and anxious officials alike to contemplate admission of bird species previously ruled out as too risky. From the earliest days of US territorial rule in Hawai'i, agriculture and forestry experts had prioritized forest preservation and health in order to protect the watersheds on which Hawaiian agriculture depended, as part of a well-established tradition in imperial discourses that associated deforestation with desiccation.[54] By the mid-1920s, however, insect infestations had reached alarming proportions on Maui. In late April 1926, Harry A. Baldwin, a member of the powerful sugar family that had founded the "Big Five" corporation of Alexander and Baldwin, urged the BCAF to bring in insectivorous forest birds to deal with koa forests on Maui beset with caterpillar and spider mite infestations.[55] Days later, David T. Fleming, a ranch and pineapple plantation manager and nature enthusiast on Maui, echoed those sentiments and pressured the board's chairman, George I. Brown, to take necessary action "to check the ravages of the various insects now wrecking our native Hawaiian Forests which in some form must be maintained at all costs as a protection for water-sheds." Woodpeckers, he argued, "are the birds which are most likely to do the greatest good."[56]

These proposals, which ran against established practices, signaled the seriousness of the problem and the need for extreme—even desperate—measures. Agricultural officials in the territory had previously avoided bringing in birds that might adapt to Hawaiian forests. As Judd had written to the board a decade earlier, "It would not be desirable to have [introduced birds] retreat to our native forests, which now contain insects almost all of which are beneficial."[57] Woodpeckers were particularly dangerous. As R. C. L. Perkins, by then retired in England, emphasized to Judd in 1922, "I should be very careful in the matter of introducing forest birds, especially anything of the woodpecker lot. . . . Though these might destroy a good many boring insects, they would be sure to destroy a lot of the beneficial ones." In particular, Perkins insisted that birds known to feed on any hymenoptera whatsoever needed to be absolutely prohibited, given the importance of parasitic wasps to biological pest control. Such birds, he stressed, "would be too risky."[58]

Now, however, with the koa forests of Maui in "very distressing" shape, agricultural officials found themselves contemplating birds that they had always rejected earlier.[59] Ehrhorn suggested introducing the downy woodpecker and the hairy woodpecker, two California species that fed mainly on caterpillars or wood-boring insects and that posed little threat to fruit or vegetation.[60] Fleming latched upon Ehrhorn's recommendation as a source of hope, albeit in a still dire situation: "Our forests are all going so fast that something must be done to protect them."[61]

The BCAF appears not to have gone so far as to bring in either of the woodpeckers, but it did quickly authorize the introduction of the Chinese thrush, or huamei, since it had already been established without apparent ill effects on Oʻahu, and Baldwin soon reported the release of fifty-two live specimens in the Olinda forest. The board seemed favorably inclined toward the Pekin nightingale (*Leiothrix lutea*) as well.[62] Discussions of these species, however, prompted some responses at length from HSPA experts that showed how a more systemic ecological consciousness was beginning to weigh into decision-making. Where protection of economic interests had previously defined responses to proposed bird introductions, scientific personnel at the HSPA now began to raise concerns about species protection and the maintenance of existing island ecosystems. As Muir warned Agee, "The addition of every animal to our fauna adds to the complexity of our biological conditions and thereby possibly adds to the increase of difficulties in biological control. For this reason I would take no risks. Unless this is taken into account no just judgment can be passed upon the introduction of any animal into our islands. In introducing insects this has been fully considered and has guided our actions, and if it should ever be ignored we shall be riding for a bad fall." Muir contended that all insectivorous birds would threaten injurious and beneficial insects alike, and that ornithologists' general observations about birds' habits could not replace direct knowledge about how introduced species would behave in a new Hawaiian habitat. "Nothing," he insisted, "except a study of the feeding habits of this bird in its native haunts, along with a knowledge of the insects of those haunts compared with the insect fauna of our islands, will allow of the most noted ornithologist forming a judgment of any value."[63]

To this established defense of experimentation and call to take introduced insect parasites into account, Muir then added an argument from "the sentimental side" that insisted on protecting Hawaiian species: "We have, in our islands, the most isolated and endemic fauna and flora in the world. So unique are they that I believe that future naturalists will condemn us if we do not take every care to preserve them intact." He added that as much as he himself loved birds, "before adding to the complexity of our biological conditions I would ask that the subject be studied from both the economic and sentimental aspects."[64]

This concept of simplified island habitats with natural relationships highly vulnerable to the introduction of new species quickly reframed discussions.

Days later, Agee echoed Muir's concerns. Biological control through the introduction of insect parasites, he argued, "has been possible owing to the very simple biologic situation which exists here," in an island habitat largely free of animals "which prey indiscriminately upon insect life." Consequently, "any birds, reptiles, or insects of a predacious nature which we introduce tend to make this simple biologic relationship more complex and thereby reduce the chances of maintaining the beneficial insects we have introduced."[65] Faced with the conundrum of weighing the urgent need to do something about the rapidly weakening koa forests against the risk that novel bird liberations "in these islands might upset the biological complexes which are now comparatively simple," Charles S. Judd appealed to the Hawaiian Entomological Society for advice.[66] He also asked the HSPA experiment station for input on how best to address the "two opposing ideas" of watershed conservation versus protection of Hawaiian ecosystems, and whether there might be safer options for dealing with the voracious caterpillars that were consuming the koa trees.[67]

This early phase of deliberations over bird introductions ended on this indefinite, indecisive note, but the discussion nonetheless highlighted the intricacies of the biological management of territory as a component of colonial rule. A soundscape devoid of birdsong drove European and American settlers' desire for a natural order familiar to them, but economic interests took priority over beautification for disciplinary experts concerned with the colonial development of agriculture and forestry. The various candidate species under consideration, especially those from California and Australia, along with possibilities from China and Japan, spoke to the Territory of Hawaii's place within a transpacific nexus tied not just by perceived similarities of climate and landscape but also by trade routes and geopolitical relations. By the mid-1920s, nascent concern with habitat and the "biological complexes" distinct to island ecosystems augmented earlier discourses about the need to defend economically valuable species and to maintain the watersheds that were critical to human survival and prosperity. Within such discussions, birds occupied an uneasy position among the competing interests of multiple other species, as both potential benefactors of cattle and koa forests through their capacity to reduce populations of invasive insects and potential threats to native Hawaiian species and to the insects that experts had introduced to defend the Territory of Hawaii's agricultural interests.

CONCLUSION

The historical episodes recounted here call attention to more than just the practical challenges of biological control, its reliance on global networks, and its relationship to the cattle trade and other economic interests that melded colonialism and capitalism in the late nineteenth and early twentieth centuries. They should also be understood as a central part of the history of state-sponsored infrastructure. No less than railroads, canals, roads, highways, dams, irrigation systems, or other technological systems associated with statist projects of economic development, the biological systems of management described here also operated as crucial human-built structures designed to maintain the basic conditions of lives and livelihoods, whether by protecting the marketability and productivity of cattle or preserving the critical functions of watersheds. A Hawaiian landscape with cattle, insect predators of horn flies from South Africa and other parts of the world, and insect-feeding birds from Australia and East Asia was no less a built environment than the technological systems that comprise the usual referents for infrastructure.

As with railroads and other infrastructural projects, the human promoters of biological control agents hoped that the species they introduced would solve problems so effectively as to fade into the background as part of the taken-for-granted conditions of normal life. Just as technological infrastructures naturalized themselves as part of expected modes of human existence, horn fly predators and insectivorous birds offered the potential to establish themselves as biological infrastructure.[68] If they succeeded, the result would be horn fly–free cattle and healthy koa forests, created by human interventions so invisible to the casual eye as to seem natural.

Biological control agents rarely achieved such unequivocal success, however. *Muscidifurax vorax* failed to make a significant dent in the horn fly population. As for the introduced avians, *Rhipidura tricolor* disappeared by the late 1930s, and more generally, introduced birds posed potential threats to native bird species while providing little in the way of effective biological control.[69] Mixed results nonetheless helped validate and perpetuate the role of agricultural experts. Animals' mobility provided constant work for the state, through a steady supply of problems to address in a world not of intermittent crises, but an endless series of challenges. The shortcomings of *Muscidifurax*, for example, merely redeployed entomologists to search for other possibilities.

Entomologists' skills remained relevant, as the key biological control strategists able to identify, collect, transport, and breed new species from distant parts of the world, as well as to facilitate cooperative relationships between experts and lay communities to establish insect sojourners in new habitats and test out their capabilities. Governmental structures also gave territorial experts primary authority to decide whether novel bird introductions would constitute a blessing or a curse for the islands, and their professional relationships with entomologists and ornithologists elsewhere in the territory or around the world facilitated the disciplinary ties necessary to uphold and extend their purview.

Ultimately, ecological and professional relationships helped define the Territory of Hawaii's place within the larger framework of global imperial order. Tropical agriculture was a discipline born of empire, and notions of shared tropicality created ties between distant places.[70] These connections were not only ideological but material in form, and defined by the living organisms that established themselves in novel habitats. The interimperial networks that brought *Muscidifurax vorax* from South Africa to Hawai'i, or birds from Australia to the islands, were part of a far-reaching interimperial nexus. Linkages between distant places defined the flows of plants and animals, as well as the scientific and political relationships that went with them. Insects, their parasites, and birds with the potential to beautify landscapes or control insect pests participated in the human history of empire through their roles as living creatures that established roots in new places. Their presence simultaneously registered and defied human ordering, through uncontrollable proclivities that underscored both imperial aspiration and its limits.

NOTES

1. W. G. Ogg to Mssrs. C. Brewer & Co., Ltd., July 16, 1913, Box 11, Folder "W. M. Giffard: Hornfly Parasite—Distribution July–December 1913," Records of the Board of Commissioners of Agriculture and Forestry, Hawai'i State Archives, Honolulu. Collection cited hereafter as BCAF.

2. D. T. Fullaway, "The Horn-Fly Problem," n.d. (July 1919?), Box 14, Folder "Division of Entomology, Foreign Explorations: Fig Insect—India, China, Malaya, August 28, 1920–May 19, 1921," BCAF. This memo and other material cited from this folder appear to have been misfiled in the archival holdings of the Hawai'i State Archives. Alternatively, if my own records are in error, this document and other

material may be in Box 14, Folder "Division of Entomology, Foreign Explorations: Horn Fly Parasite—Southwestern U.S., Mexico, July 2, 1919–February 22, 1922," BCAF.

3. John Ryan Fischer, *Cattle Colonialism: An Environmental History of the Conquest of California and Hawai'i* (Chapel Hill: University of North Carolina Press, 2015), 38–39, 55; David A. Chang, *The World and All the Things Upon It: Native Hawaiian Geographies of Exploration* (Minneapolis: University of Minnesota Press, 2016), 5; Jordan Sand, "People, Animals, and Island Encounters: A Pig's History of the Pacific," *Journal of Global History,* November 3, 2021, 1–19 (online preprint).

4. Fischer, *Cattle Colonialism*, 18–22, 30–35; Jeffrey T. Foster, "The History and Impact of Introduced Birds," in *Conservation Biology of Hawaiian Forest Birds: Implications for Island Avifauna*, ed. Thane K. Pratt, Carter T. Atkinson, Paul C. Banko, James D. Jacobi, and Bethany L. Woodworth (New Haven, CT: Yale University Press, 2009), 312, 325.

5. Edward Deveson, "Parasites, Politics, and Public Science: The Promotion of Biological Control in Western Australia, 1900–1910," *British Journal for the History of Science* 49 (June 2016): 232, 234–35; Stuart McCook, *Coffee Is Not Forever: A Global History of the Coffee Leaf Rust* (Athens: Ohio University Press, 2019), 13–14, 40–42, 66–69; Barrie Ryne Blatchford, "Pest Panic in the American West: The San Jose Scale as Change Agent in American Agriculture, 1880–1900," MA thesis, University of British Columbia, 2017; *Laws of His Majesty Kalakaua I. King of the Hawaiian Islands, Passed by the Legislative Assembly at Its Session 1888* (Honolulu: Gazette Publishing, 1888), 73; "An Act Relating to the Suppression of Plant Diseases, Blight, and Insect Pests," *Laws of His Majesty Kalakaua I. King of the Hawaiian Islands, Passed by the Legislative Assembly at Its Session 1890* (Honolulu: Gazette Publishing, 1890), 4–7.

6. Michael W. DuPonte and Linda Burham Larish, "Horn Fly," Cooperative Extension Service, College of Tropical Agriculture and Human Resources, University of Hawai'i, Mānoa, September 2003, https://www.ctahr.hawaii.edu/oc/freepubs/pdf/LM-10–11.pdf; Albert Koebele, "Report of the Entomologist," in *Report of the Minister of the Interior to the President of the Republic of Hawaii for the Year Ending December 31, 1898* (Honolulu: Hawaiian Gazette, 1899), 84–85, quotation on 85. Hawai'i had large cattle herds by the mid-nineteenth century, so it is difficult to believe that horn flies would have gone unnoticed had they arrived with Captain Cook or other early European explorers. By contrast, Koebele and other agricultural experts carefully documented the horn fly's path of migration in the 1880s and 1890s.

7. Richard C. Sawyer, "Monopolizing the Insect Trade: Biological Control in the USDA, 1888–1951," *Agricultural History* 64 (Spring 1990): 273–74.

8. O. H. Swezey, "Biographical Sketch of the Work of Albert Koebele in Hawaii," *Hawaiian Planters' Record* 29, no. 4 (October 1925): 364–68; *Second Report of the Board of Commissioners of Agriculture and Forestry of the Territory of Hawaii for the Year Ending December 31, 1905* (Honolulu: Hawaiian Gazette, 1906), 114, 131; *Third Report of the Board of Commissioners of Agriculture and Forestry of the Territory of Hawaii for the Year Ending December 31, 1906* (Honolulu: Hawaiian Gazette, 1907), 147–49; *Fifth Report of the Board of Commissioners of Agriculture and Forestry of the Territory of Hawaii for the Year Ending December 31, 1908* (Honolulu: Hawaiian Gazette, 1909), 119–22; *Report of the Board of Commissioners of Agriculture and Forestry of the Territory of Hawaii for the Biennial Period Ending December 31st, 1910* (Honolulu: Hawaiian Gazette, 1911), 116–20, 143–45; *Report of the Board of Commissioners of Agriculture and Forestry of the Territory of Hawaii for the Biennial Period Ending December 31st, 1912* (Honolulu: Honolulu Star-Bulletin, 1913), 145–49.

9. I have written about Silvestri's expedition in a separate unpublished manuscript: Jessica Wang, "The Mediterranean Fruit Fly in the Global Arena: Biological Pest Control, Inter-Imperial Relations, and American Empire in Early Twentieth Century Hawai'i."

10. F. Silvestri, *Report of an Expedition to Africa in Search of the Natural Enemies of Fruit Flies* (Honolulu, 1914), 39.

11. Appendix, Table "Daily Record of Breedings of Muscidifurax Vorax, 1913," *Report of an Expedition to Africa in Search of the Natural Enemies of Fruit Flies*, 163 (for numbers of parasites bred on a daily basis, June to December 1913); and Appendix, D. T. Fullaway, "Report for the Period from May 16 to September 30," *Report of an Expedition to Africa in Search of the Natural Enemies of Fruit Flies*, 151 (for the quotation).

12. In the case of the fungal coffee leaf rust, Stuart McCook has pointed out that steamships and the ever-increasing pace of late nineteenth-century transportation networks facilitated the rust's explosive spread in the 1880s and 1890s. In earlier time periods, long ocean voyages and shipboard conditions established high barriers for anthropogenic transfers of the rust. McCook's account underscores the need to think carefully about the precise circumstances that facilitate species' mobility, and it calls attention to the late nineteenth century as a significant era in the global history of invasive species. McCook, *Coffee Is Not Forever*, 66–69.

13. W. M. Giffard to J. M. Dowsett, A. Waterhouse, A. H. Rice, and H. M. von Holt, September 11, 1913, Box 10, Folder "Communications 1912–1914," BCAF.

14. W. M. Giffard to L. von Tempsky, August 6, 1913, Box 11, Folder "W. M. Giffard: Hornfly Parasite—Distribution July–December 1913," BCAF.

15. Appendix, D. T. Fullaway, "Report for the Period from May 16 to September 30," *Report of an Expedition to Africa in Search of the Natural Enemies of Fruit Flies*, 152.

16. See, for a typical example, W. M. Giffard to C. M. Cooke, Limited, August 7, 1913, Box 11, Folder "W. M. Giffard: Hornfly Parasite—Distribution July–December 1913," BCAF.

17. W. M. Giffard to L. von Tempsky, August 6, 1913; E. M. Ehrhorn to J. H. Raymond, August 31, 1914 [1913]; W. M. Giffard to C. H. Judd, September 3, 1913; all in Box 11, Folder "W. M. Giffard: Hornfly Parasite—Distribution July–December 1913," BCAF.

18. W. M. Giffard to Palmer P. Woods, September 8, 1913, Box 11, Folder "W. M. Giffard: Hornfly Parasite—Distribution July–December 1913," BCAF.

19. J. D. McVeigh to E. M. Ehrhorn, April 22, 1914, and Emil van Lil to J. D. McVeigh, April 2, 1914, both in Box 30, Folder "Board of Agriculture and Forestry[,] Division of Entomology[,] Hornfly Parasites[,] August 1913–January 1915," BCAF.

20. Emil van Lil to E. M. Ehrhorn, May 22, 1914; E. M. Ehrhorn to Emil Van Lil, May 28, 1914; and Emil Van Lil to E. M. Ehrhorn, June 12, 1914 (for the quotation), all in Box 30, Folder "Board of Agriculture and Forestry[,] Division of Entomology[,] Hornfly Parasites[,] August 1913–January 1915," BCAF.

21. [Illegible], manager, Molokai Ranch to BCAF, August 16, 1913, and W. G. Ogg to W. M. Giffard, August 28, 1913, both in Box 11, Folder "W. M. Giffard: Hornfly Parasite—Distribution July–December 1913," BCAF.

22. A. Mason to W. M. Giffard, December 5, 1913, and W. M. Giffard to A. Mason, December 8, 1913, both in Box 11, Folder "W. M. Giffard: Hornfly Parasite—Distribution July–December 1913," BCAF.

23. See, for example, D. T. Fullaway to W. G. Ogg, and D. T. Fullaway to Frederick Meyer, August 23, 1913, both in Box 11, Folder "W. M. Giffard: Hornfly Parasite—Distribution July–December 1913," BCAF.

24. [Illegible] to W. M. Giffard, August 26, 1913, Box 11, Folder "W. M. Giffard: Hornfly Parasite—Distribution July–December 1913," BCAF.

25. Mrs. Knott to David T. Fullaway, August 9, 1913, Box 30, Folder "Board of Agriculture and Forestry[,] Division of Entomology[,] Hornfly Parasites[,] August 1913–January 1915," BCAF; A. Koebele, "The Horn Fly," in *Third Report of the Board of Commissioners of Agriculture and Forestry of the Territory of Hawaii for the Year Ending December 31, 1906* (Honolulu: Hawaiian Gazette, 1907), 164.

26. Donald S. Macalister to W. M. Giffard, August 15, 1913, Box 11, Folder "W. M. Giffard: Hornfly Parasite—Distribution July–December 1913," BCAF.

27. G. C. Munro to W. M. Giffard, August 22, 1913, and G. C. Munro to David T. Fullaway, August 29, 1913, both in Box 11, Folder "W. M. Giffard: Hornfly Parasite—Distribution July–December 1913," BCAF.

28. John William Siddall, ed., *Men of Hawaii*, vol. 2, rev. (Honolulu: Honolulu Star-Bulletin, 1921), 293.

29. O. H. Swezey and D. T. Fullaway, "Walter M. Giffard: A Biographical Sketch," *Proceedings of the Hawaiian Entomological Society* 7 (April 1931): 347–51; George F. Nellist, ed., *The Story of Hawaii and Its Builders* (Honolulu: Honolulu Star-Bulletin, 1925), 465–67.

30. W. M. Giffard to W. H. Shipman, November 4, 1913, and W. M. Giffard to Palmer Woods, November 29, 1913, both in Box 11, Folder "W. M. Giffard: Hornfly Parasite—Distribution July–December 1913," BCAF.

31. Edward M. Ehrhorn to J. H. Raymond, September 8, 1914, and E. M. Ehrhorn to Hans Isenberg, January 12, 1915, both in Box 30, Folder "Board of Agriculture and Forestry[,] Division of Entomology[,] Hornfly Parasites[,] August 1913–January 1915," BCAF.

32. D. T. Fullaway, "The Horn-Fly Problem," n.d. (July 1919?), Box 14, Folder "Division of Entomology, Foreign Explorations: Fig Insect—India, China, Malaya, August 28, 1920–May 19, 1921," BCAF.

33. C. S. Judd to A. L. C. Atkinson, July 31, 1921, Box 14, Folder "Division of Entomology, Foreign Explorations: Fig Insect—India, China, Malaya, August 28, 1920–May 19, 1921," BCAF.

34. Matthew D. Evenden, "The Laborers of Nature: Economic Ornithology and the Role of Birds as Agents of Biological Pest Control in North American Agriculture, ca. 1880–1930," *Forest Conservation History* 39 (October 1995): 172–83; James E. McWilliams, "Biological Control, Transnational Exchange, and the Construction of Biological Thought in the United States, 1840–1920," in *Nation States and the Global Environment: New Approaches to International Environmental History*, ed. Erika Marie Bsumek, David Kinkela, and Mark Atwood Lawrence (New York: Oxford University Press, 2013), 172–73; Deveson, "Parasites, Politics, and Public Science," 234–35, 237.

35. "(Copy) Rule and Regulation by the Board of Commissioners of Agriculture and Forestry, Concerning the Importation or Introduction into the Territory of Hawaii of Birds, Reptiles, and Insects Injurious or Detrimental to Agriculture, Horticulture, or Forestry," approved October 19, 1904, copy in Box 11, Folder "W. M. Giffard: re. Importation of Birds, May 20, 1912–April 1, 1913," BCAF.

36. E. M. Ehrhorn to W. M. Giffard, May 20, 1912, Box 11, Folder "W. M. Giffard: re. Importation of Birds, May 20, 1912–April 1, 1913," BCAF.

37. W. M. Giffard to R. C. L. Perkins, May 25, 1912, Box 11, Folder "W. M. Giffard: re. Importation of Birds, May 20, 1912–April 1, 1913," BCAF.

38. W. A. Bryan to W. M. Giffard, May 31, 1912, Box 11, Folder "W. M. Giffard: re. Importation of Birds, May 20, 1912–April 1, 1913," BCAF.

39. R. C. L. Perkins to W. M. Giffard, June 4, 1912, Box 11, Folder "W. M. Giffard: re. Importation of Birds, May 20, 1912–April 1, 1913," BCAF.

40. Committee on Entomology (W. M. Giffard, chairman) to Commissioners of the Board of Agriculture and Forestry, July 3, 1912, Box 11, Folder "W. M. Giffard: re. Importation of Birds, May 20, 1912–April 1, 1913," BCAF.

41. W. M. Giffard to W. S. Wise, July 5, 1912, Box 11, Folder "W. M. Giffard: re. Importation of Birds, May 20, 1912–April 1, 1913," BCAF.

42. William E. Wall to Walter M. Giffard, February 17, 1913, Wm. A. Bryan to W. M. Giffard, March 3, 1913, and W. M. Giffard to Wm. S. Wise, April 1, 1913, all in Folder "W. M. Giffard: re. Importation of Birds, May 20, 1912–April 1, 1913," BCAF.

43. R. S. Hosmer to H. W. Henshaw, April 29, 1913, Folder "W. M. Giffard: re. Importation of Birds, May 20, 1912–April 1, 1913," BCAF.

44. E. W. Nelson to Edward M. Ehrhorn, August 20, 1919, Box 14, Folder "Birds—General, June 12, 1917–April 28, 1927," BCAF.

45. "James Franklin Illingworth in Far North Queensland, 1917–1921: 'Useful Birds' and an Overlooked Record of Insectivory in the Australasian Figbird, *Sphecotheres vieilloti*," *North Queensland Naturalist* 52 (2022): 3, 7; Ian Tyrrell, *True Gardens of the Gods: Californian-Australian Environmental Reform, 1860–1930* (Berkeley: University of California Press, 1999) (on the California-Australia horticultural nexus). On p. 15, Tyrrell notes the broader transpacific connections at work beyond California and Australia.

46. J. F. Illingworth to David T. Fullaway, June 6, 1921, Box 14, Folder "Birds—Importation of, May 25, 1915–July 11, 1927," BCAF.

47. Copy, J. Grinvell [*sic*] to D. T. Fullaway, July 7, 1921 (for the quotation) and D. T. Fullaway to Commissioners of Agriculture and Forestry, July 15, 1921, both in Box 14, Folder "Birds—Importation of, May 25, 1915–July 11, 1927," BCAF.

48. Fred Muir to the Director, Experiment Station, HSPA, November 10, 1921, and H. P. Agee to J. N. S. Williams, November 21, 1921, both in Box 14, Folder "Birds—Importation of, May 25, 1915–July 11, 1927," BCAF; and on Muir, Emmett R. Easton, "Exploits of Some Famous Entomologists of the Hawaiian Entomological Society," *Proceedings of the Hawaiian Entomological Society* 39 (2007): 153–56.

49. Foster, "The History and Impact of Introduced Birds," 312.

50. Fred Muir to the Director, Experiment Station, HSPA, November 10, 1921, Box 14, Folder "Birds—Importation of, May 25, 1915–July 11, 1927," BCAF.

51. "The Magpie Lark" (two separate documents with this heading), attached to D. T. Fullaway to BCAF, May 11, 1922, and Edw. M. Ehrhorn to Chas. S. Judd, September 10, 1924, both in Box 14, Folder "Birds—Importation of, May 25, 1915–July 11, 1927," BCAF.

52. C. S. Judd to HSPA Experiment Station, February 6, 1922; cablegram, C. E. Pemberton to HSPA Experiment Station, February 20, 1922; George C. Fuller, Premier, New South Wales, to W. R. Farrington, Governor, Territory of Hawaii, May 30, 1922; Edward M. Ehrhorn to George I. Brown, March 11, 1926; all in Box 14, Folder "Birds—Importation of, May 25, 1915–July 11, 1927," BCAF. Note the level of diplomacy at work, as symbolized by communications between the highest ranking governmental officials, namely, Hawai'i's territorial governor and the premier (equivalent of governor) in the Australian state of New South Wales.

53. C. S. Judd to E. M. Ehrhorn, August 25, 1922 and C. S. Judd to H. A. Baldwin, May 11, 1926, both in Box 14, Folder "Birds—Importation of, May 25, 1915–July 11, 1927," BCAF.

54. See, for example, Territory of Hawaii, *Report of the Commissioner of Agriculture and Forestry for the Year Ending December 31st 1900* (Honolulu: Hawaiian Gazette, 1901), 14–15. It should also be noted that tree species such as koa or māmane are found only in the Hawaiian Islands, although species preservation for its own sake was not the foremost concern in the BCAF's approach to forest management. On deforestation and colonial fears of desiccation from the early modern period onward, see Richard H. Grove, *Green Imperialism: Cultural Expansion, Tropical Island Edens and the Origins of Environmentalism, 1600–1860* (Cambridge: Cambridge University Press, 1995). Diana K. Davis has emphasized the ideological nature of desiccationist discourses and cast doubt about their scientific validity in northern Africa in light of more recent ecological evidence. See Davis, *Resurrecting the Granary of Rome: Environmental History and French Colonial Expansion in North Africa* (Athens: Ohio University Press, 2007).

55. Harry A. Baldwin to Chas. S. Judd, April 27, 1926, Box 14, Folder "Birds—Importation of, May 25, 1915–July 11, 1927," BCAF.

56. D. T. Fleming to George I . Brown, May 5, 1926, Box 14, Folder "Birds—Importation of, May 25, 1915–July 11, 1927," BCAF.

57. C. S. Judd to BCAF, February 15, 1915, Box 14, Folder "Birds—Importation of, May 25, 1915–July 11, 1927," BCAF.

58. R. C. L. Perkins to C. S. Judd, July 14, 1922, Box 14, Folder "Birds—Importation of, May 25, 1915–July 11, 1927," BCAF.

59. C. S. Judd to Harry A. Baldwin, April 30, 1926, Box 14, Folder "Birds—Importation of, May 25, 1915–July 11, 1927," BCAF.

60. Edw. M. Ehrhorn to Geo. I. Brown, May 7, 1926, Box 14, Folder "Birds—Importation of, May 25, 1915–July 11, 1927," BCAF.

61. D. T. Fleming to George I. Brown, May 11, 1926, Box 14, Folder "Birds—Importation of, May 25, 1915–July 11, 1927," BCAF.

62. C. S. Judd to H. P. Agee, May 11, 1926 and Harry A. Baldwin to C. S. Judd, May 15, 1926, Box 14, Folder "Birds—Importation of, May 25, 1915–July 11, 1927," BCAF.

63. F. Muir to the Director, May 20, 1926, Box 14, Folder "Birds—Importation of, May 25, 1915–July 11, 1927," BCAF.

64. F. Muir to the Director, May 20, 1926, Box 14, Folder "Birds—Importation of, May 25, 1915–July 11, 1927," BCAF.

65. H. P. Agee, "Memorandum Regarding Bird Introductions," May 26, 1926, Box 14, Folder "Birds—Importation of, May 25, 1915–July 11, 1927," BCAF.

66. C. S. Judd to Hawaiian Entomological Society, May 28, 1926, Box 14, Folder "Birds—Importation of, May 25, 1915–July 11, 1927," BCAF.

67. C. S. Judd to Experiment Station, HSPA, May 29, 1926, Box 14, Folder "Birds—Importation of, May 25, 1915–July 11, 1927," BCAF.

68. I derive this notion from William Cronon's conception of second nature as the built technological infrastructure that not only erased a nonhuman first nature but also rendered new ways of being natural and inevitable in the process. Cronon, *Nature's Metropolis: Chicago and the Great West* (New York: W. W. Norton, 1991), esp. ch. 2. The problematic aspects of Cronon's first nature / second nature distinction, including the too-easy assumption of first nature's nonhuman character, does not negate the usefulness of his insights about infrastructure.

69. George C. Munro, *Birds of Hawaii* (Honolulu: Tongo Publishing, 1944), 81 (on the last known sighting of *Rhipidura tricolor* in 1937); Robert L. Pyle and Peter Pyle, "The Birds of the Hawaiian Islands: Occurrence, History, Distribution, and Status," version 2, January 1, 2017, "Non-Established Species List," http://hbs.bishopmuseum.org/birds/rlp-monograph/NonEstablished.htm (June 2022); Foster, "The History and Impact of Introduced Birds," 317, 322, and 324.

70. Deborah J. Neill, *Networks in Tropical Medicine: Internationalism, Colonialism, and the Rise of a Medical Specialty, 1890–1930* (Stanford, CA: Stanford University Press, 2012), 64.

ANDREA RINGER

six Captive Breeding and the Commodification of "Surplus" Animals at the Central Park Zoo, 1886–1974

Miss Murphy, a renowned hippopotamus mother, first entered the purview of US zoo enthusiasts in 1886, when the Central Park Zoo purchased her from Carl Hagenbeck, a well-known animal trapper and trader, and arranged for her transport across the Atlantic. She arrived with the name Fatima, a reference to her capture less than three years earlier, and likely a reference to the recent Mahdist War, which had blocked the animal trade business out of the Sudan, resulting in fewer animals on the market.[1] But Fritz Koester, Hagenbeck's employee who accompanied Fatima on the SS Elder, referred to her as "Murphy" during the transatlantic journey, which reportedly elicited tail wags from the young hippo. The nickname stuck. As one of the few hippos in the United States, and with personality traits deemed "gentle and docile," she stood to be a popular attraction.[2] She was a novelty at the Central Park Zoo, which was otherwise not known as one of the major zoological institutions in the United States.

When she gave birth, Miss Murphy became even rarer among captive hippos. The Central Park Zoo gave several indications that they hoped for a successful hippo-breeding program, most notably when they purchased Caliph, a male hippo from the Cincinnati Zoo, to share the enclosure. Miss Murphy's first baby eight months later shocked Central Park zookeepers. They had been unaware of their female hippo's pregnancy and likely had little knowledge of gestation periods. Given Miss Murphy's reported age when she was captured, this was certainly her first calf. The first time a zookeeper engaged with the new pair, anxiety seemed to ensue for both mother and baby: "Sight of the man shook [the calf's] nervous system to the roots, for it ran with a frightened squeal, splashed into the water, swam to the far side of Miss Murphy, and, with chin on the mother's broad back, gazed frightened at the intruder. Then Miss Murphy, never before known to offer violence to anyone, came lumbering out of the tank, mouth wide open, showing the short, yellow tusks, and advanced to run down the enemy with her head and to grind him in her powerful maw."[3]

Although zookeepers noted that they had not previously witnessed this sort of behavior from Miss Murphy, it was likely not completely unexpected. Her direct confrontation with the zookeeper appeared to closely align with other accounts of mother hippopotamuses with their calves. The animal trade industry, which stocked zoos, circuses, and other spaces of spectacle at the turn of the century, confronted hippo motherhood during each wild capture, where mothers had garnered a fierce reputation. These interactions formed the entirety of hippo mother and infant knowledge prior to late nineteenth-century novelty breeding programs.

Like her uncaged counterparts in the Nile River in the late nineteenth century, Miss Murphy could not keep her calf from the threat of human intervention. Miss Murphy lost the infant within a few days when zookeepers snuck into the enclosure to whisk him into an impromptu nursery. This interaction was an anomaly, but it also likely had an impact on Miss Murphy's mothering behavior with future calves. Despite taking the newborn, settling him in a warmer indoor climate, and placing him on a bottle, "within ten days the baby died—a martyr to science."[4] This seemingly cold outlook on the infant's death encapsulated the purpose of the hippo-breeding program at the Central Park Zoo. As the first hippo infant born in a US zoo, with only a handful of counterparts born previously in European zoos, his value to the animal spectacle industry rested on more than his survival. The captive-breeding program operated on a trial-and-error basis, with infant deaths both expected and insightful for zoological institutes. When the Central Park Zoo and other institutions embarked on captive-breeding programs, they set out with limited knowledge sourced almost exclusively from thirdhand stories of brief encounters with wild animals. Hunters often recounted versions of a warning that "no beast guards its young as the hippopotamus does."[5] But these observations did not inform zoos of appropriate care needed for an infant to survive, like water depth and temperature for newborns. Unbeknownst to zookeepers, baby hippos did not naturally thrive in the cold and shallow waters provided in the Central Park Zoo, where they went to nurse.

Miss Murphy and her breeding partner, Caliph, each captured in Africa and brought across the Atlantic, served as the reproductive seeds to generations of captive hippos, eventually became postmortem displays in museums, and provided a bedrock of hippo reproductive knowledge.[6] Through these lived experiences as commodifiable beings from the moment of capture, hippos became the "living capital" that Josh Kercsmar refers to in chapter 1. Miss Murphy's experiences as a captive animal mother in the late nineteenth century mirrored those of other animals around the world as zoos and circuses cropped up with the rise of the animal importation industry connected to colonization.[7] Yet as the first successful hippo-breeding program in the United States, Miss Murphy and her infants present a unique lens to view the institutional decision to rely on captive rather than wild reproductive labor.

These messy transitions from wild to captive, and subsequent separations between mothers and infants, blur the line between spectacle and conservation missions in the early twentieth century. Even as most infants over the following decades could not be immediately removed from Miss Murphy due to her ferocity, the loss of Miss Murphy's first calf served as a blueprint for the success of the hippopotamus breeding program at the Central Park Zoo as it mapped out early approaches to ex situ conservation and its implications. The program's lucrative nature, which funded nearly the entire zoo for decades, rested not on re-creating the spectacle of a natural bloat, a social group of hippos, but instead on the near-constant display of mother and infant, as well as selling and shipping unwanted older calves.

On a larger scale, the early hippo-breeding program at the Central Park Zoo demonstrated that to financially sustain themselves, zoological institutes inevitably created surplus populations of certain species that were never designed to be sustained at that institute. In this way, Miss Murphy's reproductive labor allowed the Central Park Zoo to carefully curate its captive collection for the first time through culling or selling unwanted animals.[8] In the late nineteenth century, for breeding pairs like Miss Murphy and Caliph, a plethora of spectacle markets existed for young exotic animals. Their calves became "scattered . . . all over the world" as circus performers, main attractions at other zoos, and museum displays after their deaths.[9] But even with a declining animal spectacle industry by the mid-twentieth century, zoos still faced the same question of where to offload surplus animals. The breeding program of the famous first family of hippos acted as an early model for how to successfully market and implement the sale of animals under the guise of conservation. This chapter traces the opening years of captive-breeding programs through the lives of Miss Murphy, Caliph, and their calves, beginning in the 1890s until the zoo sold its final hippo in the family line in the 1970s. The rarity in both securing hippos from the wild and successfully breeding them in captivity created a niche market within the larger exotic animal trade, shifting the labor of both animals and people as infants in the zoos increasingly came under the care of human women.[10] Jennifer Marks observes in chapter 2 that the relationships between humans and animals in intertwined labor systems are intricate and usually uneven. This chapter begins to consider the intertwined feminized labor of humans and the reproductive labor of captive animals in the shift to a female workforce in the second half of the twentieth century.

Caliph and Miss Murphy's fame became intertwined as a successful hippo breeding pair in the late nineteenth century. But these hippos had nearly reached full adult maturity by the time each came to the Central Park Zoo. Their backstories are integral to understanding these hippos as historical actors. Their lives prior to zoo life mirrored each other because it was a journey experienced by nearly every captive hippo prior to the uptick of successful captive-breeding programs. As with other animals who underwent movement in the animal trade industry, the early history of Miss Murphy and Caliph is documented mostly through hearsay and rumor, although generalized adventure stories of hippo hunts provide an indication of their lives before capture.[11] But drawing from zoo and museum records, parks department reports, and newspapers, this history demonstrates how the lived experiences of captive animals remains present in institutional materials. Central Park officially purchased Miss Murphy from Hagenbeck on October 16, 1886.[12] Newspapers reported details about Miss Murphy's capture when she arrived at the Central Park Zoo. Although they were likely generalized, they were just as likely not exaggerated.

People who participated in the animal trade spoke openly about the methods and means for capturing wild animals. Miss Murphy experienced a capture that would have been familiar to other captive hippos. Hagenbeck spoke of "hippopotami pitfalls" that worked especially well since mothers sent their young ahead of them so they could scope out danger from behind. The pitfall was a hole in the ground covered in branches that an infant would fall into, leaving the mother to flee, perhaps confused and scared.[13] But outwitting a hippopotamus mother likely occurred less often than outright slaughter, which left mothers dead and young captured. Hunters typically followed tried-and-true methods to capture the now orphaned infants by harpooning them in the "fleshy part of the hindquarters" to ensure capture without any internal injuries.[14] This reliable method of capture left most captive hippos, including Miss Murphy and Caliph, with recognizable permanent scars.

Hagenbeck and his network of workers frequented areas around the Nile, where Miss Murphy and Caliph were likely born. In his adventure memoir, Hagenbeck noted that "the Egyptian Sudan is one of the richest and most inexhaustible sources of animal life."[15] But even with a wealth of animals to capture, the process was difficult. And despite Hagenbeck's description, the population did prove exhaustible.[16] Animal procurers noted that for some animals, like hippopotamuses, the separation between mother and baby occurred early. After describing the mother hippos, who "die hard," one hunter plainly stated, "We cannot get these babies too young to suit. One, I remember, was captured the very day it was born, and the hunters and attendants brought it up on a bottle."[17] P. T. Barnum, a museum and circus proprietor who profited from exaggerations in the press, claimed that his hippopotamus in the early 1880s was captured two minutes after its birth.[18] Although the account may have served as posturing for Barnum, early capture really did leave infants more vulnerable to trappers and mothers less capable of defending.

Caliph, who came to the United States three years prior to Miss Murphy, might have begun his life in Ethiopia. Since Cincinnati purchased him from the European animal trade market when he was approximately eighteen months old, his transatlantic stop in Hoboken, New Jersey, was immediately followed by a cross-country ride in a large pine box on the United States Express Company train. Although many animals crossing the Atlantic reportedly suffered from seasickness, including Miss Murphy, Caliph seemed to fare better. Once at the Cincinnati Zoo, Caliph could often be seen munching hay in his small enclosure near the wading pool. Zoogoers had to go to the Carnivora house to visit him, across from the lion cages. Early reports noted that he seemed "leisurely" and "sedate." But it was his status as a young calf that particularly attracted zoogoers to his enclosure. Having a baby hippo reportedly made more renowned zoological institutes like Philadelphia jealous.[19]

The Cincinnati Zoo regularly shopped at circus and animal trade auctions to add to their animal collection, which continued to blur the line between education and spectacle. Even when the superintendent went to Hoboken to pick up Caliph, he picked up nearly a dozen additional animals, including monkeys, lemurs, and a kangaroo.[20] While some US zoos, like the Philadelphia Zoological Garden, aimed at more scientific endeavors that resembled the London or Amsterdam Zoo, small institutions found it impossible to separate themselves from lowbrow traveling animal shows. When circuses went under, the market flooded with animals and circus equipment. Circuses infamously enlarged their outfits this way, but so did zoos. They purchased everything from boa constrictors to lion cages at affordable prices from out-of-luck circuses.[21] Central Park's success was particularly intertwined with the circus, as their menagerie took shape when P. T. Barnum used the facilities for animals in his own shows.[22]

Line items in zoo budgets demonstrated the financial investment such institutions were willing to make and their hope for a payout. Hippopotamuses regularly appeared on Hagenbeck's sale lists as the most expensive items.[23] Hagenbeck seemed to understand their market value when he sent an expedition to hunt exclusively for baby hippos in the early 1870s. Adding to their value was the difficulty in capturing them; the capture party came back empty-handed.[24] Despite the impressively high price tags that zoos bragged about paying, they often ran themselves into debt to make purchases. Three years after purchasing Caliph, the Cincinnati Zoo was still making payments toward the $5,000 debt.[25] Relative ignorance about hippo care and probable financial constraints meant that the Cincinnati Zoo had not designed an adequate enclosure for its new hippo. Though undated, a nineteenth-century postcard from the Cincinnati Zoo shows a hippo standing in water that barely reaches its torso with a fence butted up to the water on three sides. The image corroborates descriptions of Caliph's enclosure and almost certainly is a painted representation of him. The enclosure's small size and shallow water indicate the dearth in knowledge from the zoos that kept Caliph, Miss Murphy, and other captive hippos in the nineteenth century.[26]

When Caliph arrived at the Central Park Zoo eighteen months later, he joined an already popular hippo exhibit containing Miss Murphy. Park attendance increased with the new animal attractions, and a reported hundred thousand visitors entered the park on peak attendance days.[27] Citing the new animal acquisitions, including Caliph, the board claimed that "this Department has come to attract largely increased attention both at home and abroad," which proved it was "meeting a popular want."[28] The zoo immediately updated the hippo summer quarters. Less than two months after Caliph's arrival, the hippo pair entered their new enclosure, which included a 24,000-gallon tank and new railings and bars.[29]

Until Caliph and Miss Murphy began sharing an enclosure, US zoogoers were exposed to just one hippo at a time. A single hippo occupying a zoo exhibit or pulling a Ringling Brothers cart was surely thrilling, but it did not display some of the species-typical behaviors, like eating and bathing, that zoogoers sought. Animal shows and interactions remained popular at zoos in the late nineteenth century, but so did feeding sessions open to the public. Memoirs, newspaper series, drawings, photographs, and later, film appeared in the American cultural landscape and provided people with some familiarity with hippos even before they saw one in person.[30] In the 1880s, most people strolling through the Central Park Zoo would have seen drawings of hippos that depicted ferocious attacks on people venturing into the water.[31] Yet the likely first photograph of a captive hippo displayed the opposite. Taken at the London Zoo in 1852, the image shows a large hippo, Obaysch, basking in the sun beside his pool with zoogoers mere feet away.[32]

Hippos immediately garnered contradictory identities and qualities that would be exploited by spaces of animal spectacle. They were lazy, ferocious, slow, sneaky, conniving, protective, seemingly smiling, and always hungry. These contradictions emerged because captivity could not replicate the natural environment for hippos. The reaction to each of these behaviors when first exhibited by Caliph, Miss Murphy, and their infants demonstrated how public knowledge of hippo behavior formed during interactions and observations at the Central Park Zoo exhibit. The seemingly lazy hippos lounging by the pool at noon were nocturnal, and the loaves of bread that zookeepers shoveled into their mouths did not resemble the nutritious and filling diets they would have consumed in the wild with their near-constant grazing. But even if zoos could not give zoogoers the feel of being on the Nile, they still offered a degree of animal culture. For zoogoers, previous encounters with hippos had occurred on printed paper. Those descriptions and images had difficulty capturing the sight of blood sweat, a constant, red-colored sweat that gives hippos a pinkish hue, and their loud and unexpected sounds, ranging from grunting to roaring.

The presence of infant hippos put even more animal culture on display. For the first time, zoogoers witnessed known species-typical behaviors. Some of these behaviors, like minutes-long submersion in the pool, proved terrifying when days-old infants attempted it.[33] Others were less satisfying, like when observers who wanted full view of the infant noted that "the spectacle of Miss Murphy and her child is not. . . . edifying" as she rested with most of her body in the water, "the little one's hind legs . . . braced against its mother's side . . . with its little eyes closed and its nose just above the surface.[34]

The death of their first infant in 1889 prompted the zoo to try to start the breeding program again. Less than a year later, the Parks Department began describing the hippos in significantly more detail, as it now documented the making of the new hippo family with the birth and survival of the second infant. This pregnancy did not catch the institution off guard; it documented the gestation to the exact day.[35] The new baby, named Fatima, did not undergo the same separation as the previous infant. Instead, Fatima was weaned at nine months and interacted with zookeepers through the bars but stayed with her mother until her eventual sale. This later separation likely gave the calf a better chance at healthy bonding, nutrition, and social interaction.

With a healthy hippo infant, the zoo was finally able to lay claim to a successful breeding program, which warranted an additional financial investment. The zoo enlarged the hippo enclosure with a new building addition. As popular as the mother-calf pair was and despite the larger space, three hippos were "just one too many for the Central Park Zoo," and they passed an official resolution to sell Fatima in January 1893.[36] The anticipated arrival of a third baby, which indicated the continued success of the breeding program, also likely pushed the zoo to sell Fatima to another institution.[37]

The updated hippo enclosure was more conducive to the survival of mother-infant pairs, but the environment was still not a replica of the one Miss Murphy was born into along the Nile River, demonstrating the marked differences between captured and captive-born hippos. Captive-born hippos, which can also be understood as Anthropocene hippos, lived in simulated nature with constant human surveillance and near-constant human intervention. For these generations of captured and captive-born hippos, enclosures served as cages meant to offer guests some semblance of animal behavior (fig. 6.1).[38] For the Anthropocene hippos birthed by Miss Murphy, life never fell outside the purview of the public gaze, either in zoo cages or in circus performances. They

FIG. 6.1. The zookeeper slipping between the bars and the close proximity between the enclosure and zoogoers offer a glimpse into the public lives of Anthropocene hippos. The separation between the two hippo enclosures offered a controlled environment for the breeding program, as the male or older calves could be separated from Miss Murphy. Uncredited photograph, November 1944. Parks Photo Archive, New York City Department of Parks and Recreation.

never formed a sustainable bloat, and they were orphaned by sales between institutions.

Fatima's auction marked a watershed for the captive-breeding program at the Central Park Zoo. If the institution could get the price asked for, that would nearly return their original investment in Miss Murphy. But the first auction ended without a single bidder. The $3,000 starting bid seemed on par with market prices, and the sale seemed safer without the unknowns of a transatlantic journey, but the price was likely still high for most institutions. Even with representatives from James Bailey and Walter Main's circuses, both large and wealthy, there was still no bid.[39] Instead, the young hippo stayed with Miss Murphy and Caliph until she eventually gave birth to her own calf in 1896. Decades before the meticulous tracking of breeding and genetics through stud books and oversight through accreditation agencies, the birth of Fatima's infant sired by Caliph demonstrated the messy ways early breeding programs operated. Nonetheless, the zoo celebrated the new baby. Fatima gave birth during the zoo's admission hours, which prompted a constant guard to keep the gathering crowd from the enclosure.[40] When Miss Murphy also gave birth less than six months later, the zoo clearly had more hippopotamuses than it could handle.[41] Like the first baby born at the Central Park Zoo, Miss Murphy's newborn was removed from the tank and placed in a nursery.[42]

With five hippos in 1897, zoo officials reached back to their original plan to sell Fatima. This time, they received several bids but traded her to the Ringling Brothers in exchange for an elephant and two lions.[43] While Fatima traveled with the Ringling Brothers, the remaining hippos at the Central Park Zoo acclimated to the new family dynamics. Fatima's infant, Sirus, was one year old when his mother was sold to the Ringling Brothers. Although old enough to be weaned, he was still clearly nursing up until Fatima's sale. The move to the summer quarters, an outdoor facility, less than a month after the sale put Sirus in the same tank as Miss Murphy and Iris, her six-month-old calf, for the first time. But after Sirus pushed the smaller infant away from Miss Murphy to nurse, keepers had to relocate him to Caliph's tank as a temporary measure.[44] Like his mother before him, he was traded to another institution in exchange for "a lioness and other zoological specimens" while zookeepers awaited the impending birth of Miss Murphy's next infant.[45] The sale of Sirus, and then Iris six months later, to Hagenbeck's firm demonstrated the multifaceted options for hippo sales. Central Park was not constrained to US zoos to offload their surplus hippos.[46] Rather than eliciting public outcry, these sales cemented the widespread belief that animals were there to entertain and that this world was dominated by savvy businessmen and animal keepers who knew enough to keep animals alive and sell them at top-dollar prices.[47]

After the arrival of the first seven calves, who then supplied the captive population in circuses and other zoos, the press referred to the pair as the "founders of the greatest hippo family in all the world."[48] Despite the rhetorical use of "family" to describe Central Park Zoo's hippo exhibit, the continued breeding of Caliph and Miss Murphy necessitated separation of their calves. Lacking the space, resources, or desire to house the dozens of hippo calves produced through its breeding program, the zoo engaged in high-dollar sales of the young offspring, often separating them well before they reached full development.[49] The hippo exhibit at the Central Park Zoo continued to change with seasonal births and the consequent sales. Keepers watched the hippo enclosure carefully, taking shifts as Miss Murphy started to show signs of labor in 1900. Again, they took note of her newly "irritable and unpleasant" disposition.[50] That reputation only intensified as the hippo family headed to its winter quarters the following September. Keepers kept Miss Murphy and baby Lotus in a separate enclosure from Caliph. The press noted that the separation resulted from Caliph's "good-natured clumsiness," which might harm the infant.[51] But the seemingly easy trek from summer to winter quarters ended in a brawl as Caliph lunged at the infant and Miss Murphy attacked Caliph, ripping off his tusk and sending zoogoers running away screaming.[52]

Although Miss Murphy and Caliph originally came to Central Park as gentle young hippos, the ever-changing bloat of infants shifted the dynamic, and zoogoers were exposed to sparring personalities. Zoos could not control the temperament of their hippos, and the environment likely made the hippos more disgruntled. The vicious images of hippos attacking people seemed to be replicated in the zoo environment during these moments, as the dominant male in the artificial bloat at the Central Park Zoo demonstrated species-typical behavior in the wild by acting with aggression toward an infant. The zoo could no longer lay claim to hosting a herd of gentle giants. Caliph and Miss Murphy's value to the zoo became their role as reproductive seeds. It was their infants that provided gentle interactions with zookeepers and the public.

While Caliph and Miss Murphy did not have the disposition for interactive performances, their infants sometimes did. Keepers and the public seemed particularly fond of Caliph II, the last infant sired by Caliph. For the first time, keepers used hippos as the stars of the show. Unlike the seemingly unstable Miss Murphy and Caliph, their full-grown infant appeared "better natured." The press credited it as evidence that "by being born in captivity he has assimilated a lot of civilization, docility, and intelligence."[53] For the purposes of the early twentieth-century zoo, as a space where captive exotic animals could come face-to-face with zoogoers and potentially interact with humans for entertainment, this "better-natured" trait became crucial. Albert Schaad, a zookeeper, took a variety of behind-the-scenes photos in 1913 that demonstrate the deeply intertwined lives of animals and their keepers, as well as the performative nature of early twentieth-century zoos. Despite their formidable reputation and stories of near-death experiences, Schaad seemed to interact with Caliph II without any hesitation. Images showed Shaad smiling beside Caliph II with his head inside the hippo's mouth.[54]

As Miss Murphy successfully reared seven babies, the program appeared to be a resounding success. Following Caliph's death, the zoo quietly slipped Caliph II into the sire position, again compromising the genetic diversity of captive hippo populations.[55] Though the broader wild animal trade still flourished, hippos in particular were "rarely hunted" for captive display. In the wild, they were dangerous and difficult to capture alive, and the few zoo hippos already in captivity quickly showed an ability to reproduce successfully there, as demonstrated by the Central Park Zoo's experimental program.[56] The zoo also began standardizing what captive-bred wild animal sales looked like. Working from increasingly sound knowledge of hippo familial relations, the zoo held off on separating mother and calf until the calf reached two years of age and could more readily be weaned.[57] In the ensuing years, infant mortality dropped as zookeepers ensured that water depth and temperature were ideal for a mother-calf pair, and breeding became seasonal to ensure warm-weather births. They had seemingly moved on from infant deaths "on account of the menfolk not knowing what is good for such babies."[58]

The ignorant menfolk trope opened the door for women to provide seemingly more adequate care work in the ensuing decades, as the human labor involved in the mother-calf separation process had a profound effect on worker demographics. While men controlled the reproductive decisions and processes of captive animals, women increasingly moved into care work positions for separated young. Some institutions, like the Bronx Zoo, began using women's labor in unofficial capacities with the growth of infant care work and the creation of zoo nurseries in the 1930s. Official zookeeper uniforms at the Central Park Zoo, however, were only fitted for men during these early decades, perhaps indicating that zoos had no intention of hiring women in these more official capacities.[59] At the San Diego Zoo, women continued to do the same work but did so as employees. Women continued to fill opportunities for animal care work at zoos in the post–World War II years. As Daniel Bender has pointed out, most of these "mothering" roles between human women and zoo animals were part of a Cold War aesthetic of family.[60] With more successful rearing of baby animals across the institution, the Central Park Zoo found new ways to capitalize on this novel population of captive animals with the opening of the Lehman Children's Zoo in the early 1960s.[61] Headed by Tatiana

Gillette-Infante for several years, the Children's Zoo featured an array of baby animals, most without their mothers and available for interaction. Gillette-Infante served as a provisional menagerie keeper and spent much of her workday cleaning stalls and cutting meat. The press, however, fixated on the novelty of a woman doing this sort of work, noting that "she comes to work without makeup in her olive drab trousers and jacket, her long brown hair tied back in a bun." Noting that she had no children of her own, the articles remarked that "her attitude toward her [animal] charges . . . is indeed maternal."[62]

The slow climb to running a children's zoo filled with baby animals was part of a larger transition from an institution that hosted an exhaustive menagerie of animals to one that could carefully curate its collection for the public because of the financial gains from Miss Murphy's reproductive labor.[63] Central Park had engaged in a breeding program prior to Miss Murphy, but on a substantially smaller scale. In 1889, fifty-one lambs and thirty-one ewes were sold at auction, with only $605 to show for it.[64] Fifteen years later, the annual auction of surplus animals brought in similar profits.[65] These surplus animals were sold as "bargain sales," including animals like Edmee the lioness, who hailed from the "African jungle" but had a "slight limp," as well as a one-eyed hyena.[66] For the more valuable infant species bred in Miss Murphy's era, the zoo often asked not for cash but for animal trades, through which the institution acquired a collection of animals in exchange for Miss Murphy's infants. Over more than thirty years at the Central Park Zoo, Miss Murphy witnessed the menagerie of animals changing around her. The sale and trades of her young facilitated the founding of whole new departments, bringing in elephants, hyenas, and lions.[67] It created a self-supporting menagerie with a yearly surplus and no appropriated acquisitions budget.[68]

With a substantial array of animals, thanks in large part to Miss Murphy's reproductive labor, the zoo faced new ethical dilemmas associated with a large and aging population. Despite outcry from the Society for the Prevention of Cruelty to Animals (SPCA), the Central Park Zoo culled its formerly popular and increasingly elderly hyenas in 1911. Armed with rifles, keepers shot the hyenas, who were "too old and should have died years ago."[69] Age became a defining feature of the zoo's animal population as older animals were culled and healthy young animals welcomed into the collection. The zoo also had to contend with the growing carnivore population, which required fresh meat. They remedied that issue in a budget-friendly way with a proposal to purchase "old horses who had outlived their purpose and kill them ourselves, so that the meat costs us about a cent and a half a pound."[70] When Black Diamond, a buffalo acquired through a trade, reached twenty years old, the zoo made a similar decision to cull him. The zoo sold him to a butcher on East Fourteenth Street who was optimistic about the price he would get for the steaks. Buffalos, the zoo feared, could "pass away without a moment's notice" at a significant cost to the zoo, and "that wasn't good business."[71]

The zoo held such an outlook on its aging animal population because Miss Murphy's reproductive labor had set the stage for culling and selling unwanted animals either to prevent loss or for significant financial gain. Prior to the hippo-breeding program, the Central Park Zoo could not afford to offload popular animals. Instead, it desperately collected whatever it could find for its less impressive menagerie. As the zoo embarked on its hippo-selling program, a new financial approach emerged at the institution. Animals had a value on their heads, and that top dollar was not always in their caged display at Central Park. Elephants had undergone public execution for dangerous misbehavior even when zoos could not necessarily afford to replace them. Captive breeding gave zoos a larger swath of options for whether to remove an animal from its collection.[72] Whether through execution or sales, Central Park gained a greater ability to curate its animal collection.

Even after death, the bodies of the Central Park hippos continued to occupy stages.[73] Caliph's body was stuffed and placed in the New York Museum of Natural History after his death in 1910.[74] Nearly twenty years later, Miss Murphy's preserved body joined the display.[75] People could once again see the progenitors of the captive hippo population together. Some of their calves who ended up scattered on stages around the world also met the same fate. While the Milwaukee Public Museum directed expeditions to Africa designed to bring back specimens for museums, it also accepted donations. In 1909 the Ringling Brothers offered the skin and skeleton of their recently passed hippopotamus. Although the Cudahy-Massee Expedition to British East Africa brought back 263 skins, "the most valuable specimens received during the year were a fine large hippopotamus from Ringling Brothers and a young chimpanzee from the Big Otto Show, both received as gifts."[76] In a thank-you note to Al Ringling, the museum director noted a bullet beneath the shoulder blade of the deceased hippopotamus, a scar that marked the final generation of wild-born hippopotamuses.[77] Although Miss Murphy and

Caliph bore the same mark on their stuffed bodies, their calves, who would also appear in museums after death, were unmarked. This meant that every subsequent generation of Anthropocene hippos was physically, and sometimes behaviorally, different than their wild-born counterparts.

Miss Murphy's death in 1928 did not signal the end of the breeding hippo program at the Central Park Zoo, but it was the beginning of the zoo's divestment. In her final days, the zoo eulogized her reproductive value and role as a mother to more than a dozen hippos on stages around the world.[78] Although the zoo no longer held a monopoly on captive-bred hippos, the program had already left a mark on captive reproduction and the policies that surround it.

As Rosey replaced Miss Murphy for a few decades, the hippo-breeding program at the Central Park Zoo saw increased human intervention.[79] The zoo and press characterized Rosey as overly friendly to people but also a bad mother and eventually an "old maid." Her first two babies appeared unwanted, and she immediately stomped them to death.[80] Rosey's maternal behavior justified increased human intervention with infants while also marking her as an Anthropocene hippo who had less experience and interaction with calves in a bloat than her mother had, as she approached calves more carelessly. Keepers intervened with subsequent calves while the press surmised that the babies must be wondering about the whereabouts of their mother as keepers fed them on a two-hour rotation.[81] As the care work for Rosey's calves continued to be outsourced to people, the first images of women providing infant care work appeared behind the scenes, matching the clear interest from newspapers in women zookeepers (fig. 6.2).[82]

By the 1970s, breeding programs and infant care work in zoos were in full force. The hippo-breeding program at Central Park had lost both its novelty and its profit margin with Miss Murphy's death. The zoo sold its hippo family, which then consisted of Rosey, Falstaff, and Daisy, to the Toronto Zoo to make way for new pygmy hippos. According to zoo staff, these new smaller hippos were "more of a novelty . . . and they, of course, need less space than the others."[83] Of course, this smaller species of hippo also fit the growing aesthetic of babies and manageable animals. In addition to changes in the captive animal population, care work at the Central Park Zoo was also shifting. Although women had been working in the children's zoo for nearly fifteen years, Veronica Nelson became the first woman at the Central Park facility to officially serve as zookeeper.[84] Hired to care for Patty Cake, an injured infant gorilla,

FIG. 6.2. On straw-covered ground and holding a bucket of nourishment, an unnamed woman is seen feeding a ten-month-old hippopotamus calf. The image from July 1964 appears candid, rather than a publicity shot, and likely served as documentation of the work women had moved into at the Central Park Zoo. Parks Photo Archive, New York City Department of Parks and Recreation.

Nelson followed in the footsteps of women at other zoos in the early 1970s who likewise became zookeepers by specializing in leading infant care work.[85] Zoos successfully rebranded themselves as spaces of conservation when animal importation became more difficult. In contrast to the unorganized hippo-breeding program that started in the nineteenth century, the Central Park Zoo standardized its records and methods to match the formal move to associate zoos with spaces of conservation over spectacle. As the animal trade industry collapsed, zoos frantically collected the last few wild animals they would need to genetically sustain the captive population. But as the hippo-breeding program at Central Park demonstrates, this transition was not sudden. It occurred over generations, as zoos had already accumulated decades of knowledge around captive motherhood and infant survival.

Zoos were prepared to step into visible conservation roles because of novelty breeding program experience, the creation of children's zoos that perpetuated the image of humans raising animal babies, and the demographic shift to women zookeepers, who became readily associated with this public care work. The lessons learned from the reproductive labor of Miss Murphy and other captive animals at the turn of the century informed these changes. Professional organizations, such as the Association of Zoos and Aquariums, provided accreditation standards and best-practice handbooks for captive populations. Stud books that meticulously tracked the genetic diversity of captive populations emerged in the early 1960s, making the inevitable inbreeding in the early hippo program a thing of the past. The trial-and-error method that resulted in high infant mortality rates was replaced with widely available species survival plans and population viability analyses that documented minute details of motherhood.[86]

Yet surplus animals persisted. Meticulously kept details generated metrics that zoos vowed to follow, including population size. Despite apprehension from the public and some zookeepers, zoos continue to have contingency plans to cull populations.[87] Even with the decline of circuses and more robust protective legislation for animals, zoos have largely maintained the same pathways to offload surplus animals. While the legal international trade market collapsed with the US Endangered Species Act in 1973, the domestic market transformed. Miss Murphy's calves, lions with a limp, and elderly hyenas ended up in other zoos, at wild animal shows, or euthanized in the early years of the zoo. Under conservation policies nearly a century later, surplus animals were now transferred between accredited institutions, sold to stateside safaris or exotic hunting facilities, or euthanized.[88] The migration taken by Miss Murphy's calves to other zoological institutions would later be referred to as "genetic management."[89] According to Mads Frost Bertelsen, who came under heavy fire after killing Marius the giraffe at the Copenhagen Zoo in 2014, exterminating members of a captive species is "not only a sign of a healthy population, but an unavoidable 'by-product' of sustainable breeding."[90] For Bertelsen, killing the healthy giraffe who could not contribute to genetic diversity was a more viable option than possibly sending him to an unaccredited institution. Using conservation-centered language that centers the best decisions for an entire species over individuals, and despite a conscious move to distance themselves from their colonial past, zoos have maintained nearly the same range of options for their surplus animals.

NOTES

1. "Foreign Wars and the Zoo," *Tunkhannock (WY) Republican*, May 8, 1885, 4. Local armed conflict under the pretenses of colonization affected the animal trade, particularly for hippopotamuses. For a brief history of the conflict, see Gabriel Warburg, "European Travellers and Administrators in Sudan before and after the Mahdiyya," *Middle Eastern Studies* 41, no. 1 (January 2005): 55–77; Iris Seri-Hersch, "'Transborder Exchanges of People, Things, and Representations: Revisiting the Conflict between Mahdist Sudan and Christian Ethiopia, 1885–1889," *International Journal of African Historical Studies* 41, no. 1 (2010): 1–26.

2. "Fatima: The Only Available Hippopotamus in the World Reaches Central Park," *Sunday Leader* (Wilkes-Barre, PA), October 24, 1886, 30.

3. A. W. Rolker, "Babies of the Zoo," *McClure's Magazine* 21 (October 1903): 611.

4. Rolker, "Babies of the Zoo," 611.

5. "A Nubian Caravan," *The Inter-Ocean* (Chicago), June 10, 1882, 7.

6. "A Hippo's Happy Family," *Kansas City Star*, May 5, 1905, 14.

7. Scholars have explored several of these stages with respect to captive animals. Janet Davis has offered the most in-depth look at US circuses from a cultural framework. See Janet Davis, *The Circus Age: Culture and Society under the American Big Top* (Chapel Hill: University of North Carolina Press, 2003). Since the publication of Davis's book, several additional studies have emerged, forming the nascent field of circus studies. For an introduction to the field that includes animals as historical actors and critically examines networks of animal trade, see Peta Tait and Katie Lavers, eds., *The Routledge Circus Studies Reader* (New York: Routledge, 2016). Scholarship on other stages of spectacle offers broader insight into the capitalist networks of exchange under colonization. See Nigel Rothfels, *Savages and Beasts: The Birth of the Modern Zoo* (Baltimore: Johns Hopkins University Press, 2002); Susan G. Davis, *Spectacular Nature: Corporate Culture and the Sea World Experience* (Berkeley: University of California Press, 1997); Ben A. Minteer, Jane Maienschein, and James P. Collins, *The Ark and Beyond: The Evolution of Zoo and Aquarium Conservation* (Chicago: University of Chicago Press, 2018).

8. Robert Lacy, "Culling Surplus Animals for Population Management," in

Ethics on the Ark: Zoos, Animal Welfare, and Wildlife Conservation, ed. Bryan G. Norton, Michael Hutchins, Elizabeth F. Stevens, and Terry L. Maple (Washington, DC: Smithsonian Institution Press, 1995), 188. This chapter works from Lacy's definition of culling as "the termination of the life of an animal before it would have died from unavoidable disease or failures of organ systems." Lynda Birke has addressed the ethical issues of culling and its intersection with animals in laboratories. Lynda Birke, "Exploring the Boundaries: Feminism, Animals, and Science," in *Animals and Women: Feminist Theoretical Approaches*, ed. Carol J. Adams and Josephine Donovan (Durham: Duke University Press, 1995): 32–54.

9. "It Won't Be the Same Old Zoo without Bill Snyder," *Literary Digest* 57 (1918): 74.

10. This intertwined agency and opportunity between human women and female animals, which center on maternity, is embedded in an ecofeminist framework, to use a term first coined by Françoise d'Eaubonne. Some scholars leveled critiques against ecofeminism because it lacked an intersectional framework and was essentialist. More recent ecofeminist scholarship that intersects explicitly with animal studies has responded to these critiques. G. A. Bradshaw has noted that although defined in terms that can be applied in interspecies ways, motherhood is often a situational identity and process. Yet "cortical and limbic structures related to maternal and other affiliate behaviors and psychological states are shared among all mammals" (*Carnivore Minds* [New Haven, CT: Yale University Press, 2019], 60). Another recent collection of essays exploring intersections of ecofeminism has intervened in conversations about its connection to breeding, particularly among farmed animals. See Sunaura Taylor, "Interdependent Animals: A Feminist Disability Ethics-of-Care," in *Ecofeminism: Feminist Intersections with Other Animals and the Earth*, ed. Carol J. Adams and Lori Gruen (New York: Bloomsbury, 2014), 109–26.

11. Reading these stories against the grain to find the lived experiences of Caliph and Miss Murphy is part of the difficulty in finding the histories of animals in the Anthropocene. Ideas of colonialism also punctuate this history, and details of capture and intervention by European and US animal keepers idolize human roles as conquerors and caretakers of wild animals.

12. "Doctoring a Menagerie," *Hollis (NH) Times*, December 13, 1889, 7; *Minutes and Documents, Board of Commissioners, Department of Public Parks* (New York: Evening Post Printing Office, 1890): 24; Minutes and Documents of the Board of Commissioners of the Department of Public Parks, April 30. 1891, 148, New York Parks Department Historical Archives, New York City (hereafter Parks Archives).

13. Carl Hagenbeck, *Of Beasts and Men* (London: Longmans, Green, 1910), 72.

14. "A Nubian Caravan," *The Inter-Ocean* (Chicago), June 10, 1882, 3; "Catching a Live Hippo," *Newcastle News* (UK), July 14, 1909, 3.

15. Carl Hagenbeck, *Of Beasts and Men*, 47.

16. John Simons has contextualized the biography of Obaysch, a captive hippo in Europe, within a moment of increased colonization and declining wild populations. John Simons, *Obaysch: A Hippopotamus in Victorian London* (Sydney: University Sydney Press, 2019), 37.

17. "Circus Secrets and Stories," *The Inter-Ocean* (Chicago), June 16, 1901, 44.

18. "A Nubian Caravan, *The Inter-Ocean* (Chicago), June 15, 1882, 3.

19. "Caliph's Debut," *Cincinnati Enquirer*, April 23, 1883, 8.

20. "Caliph's Debut."

21. "A Circus Sold at Auction," *National Republican* (Washington, DC), December 4, 1886, 6.

22. Elizabeth Hanson, *Animal Attractions: Nature on Display in American Zoos* (Princeton, NJ: Princeton University Press, 2002), 31; Roy Rosenzweig and Elizabeth Blackman, *The Park and the People: A History of Central Park* (Ithaca, NY: Cornell University Press, 1922), 342–45.

23. Nigel Rothfels, *Of Savages and Beasts: The Birth of the Modern Zoo* (Baltimore: Johns Hopkins University Press, 2002), 217. Rothfels cites an 1883 inventory list from Hagenbeck that shows a female hippopotamus listed for 18,000 marks. The second most expensive animal, a rhinoceros, is listed for 12,000 marks.

24. Nigel Rothfels, "Catching Animals," in *Animals in Human History: The Mirror of Nature and Culture*, ed. Mary Henninger-Voss (Rochester: University of Rochester Press, 2002), 196.

25. "Annual Report of the Zoological Society of Cincinnati," 1885–86, 6, Cincinnati Public Library, Genealogy and Local History Collection.

26. "Hippopotamus at the Zoo, Cincinnati, Ohio," Paul F. Bien postcard collection, Cincinnati and Hamilton County Public Library, Genealogy and Local History Collection.

27. "Caliph the Hippo: New York Captures Him from Our Zoo for $5000," *Cincinnati Post*, April 13, 1888, 1; Minutes and Documents of the Board of Commissioners of the Department of Public Parks, April 30, 1891, 24, Parks Archives; Minutes and Documents of the Board of Commissioners of the Department of Public Parks, April 30. 1889, 9, Parks Archives.

28. Minutes and Documents of the Board of Commissioners of the Department of Public Parks, April 30. 1887, 18, Parks Archives.

29. Minutes and Documents of the Board of Commissioners of the Department of Public Parks, April 30. 1889, 8, Parks Archives.

30. "Boat Attacked by Hippopotamus," *New York Times*, August 20, 1883, 2.

31. "Dozen Hippopotami Attack Roosevelt," *New York Times*, July 22, 1909, 4; A. W. Rolker, "How Wild Beasts Are Caught for Zoos," *The Strand* 34 (1907): 515.

32. "Hippopotamus at the Zoological Gardens, Regent's Park, London," Gilman Collection, Purchase, Ann Tenenbaum and Thomas H. Lee Gift, Metropolitan Museum of Art, New York, 2005.

33. "The Babies of Wild Animals," *Democrat and Chronicle* (Rochester, NY), April 8, 1903, 13.

34. "The Biggest Mouth in New York," *New York Times*, October 30, 1890, 8.

35. Minutes and Documents of the Board of Commissioners of the Department of Public Parks, April 30. 1891, 4, Parks Archives.

36. "A Hippo Might Come in Handy," *Evening World* (New York), January 7, 1893, 5.

37. "A New Baby Hippo," *Evening World* (New York), February 3, 1893, 1.

38. Minutes and Documents of the Board of Commissioners of the Department of Public Parks, 1989, 302, Parks Archives.

39. "Work of the Park Board: Fatima, the Hippopotamus, Must Not Be Sold for Less Than $3,000," *New York Times*, January 26, 1893, 9; "Fatima Not Sold" *Evening World* (New York), January 26, 1893, 2; "No Bids for Fatima," *New York Times*, January 27, 1893, 10.

40. "The Baby Hippo's Reception Day," *New York Tribune*, March 23, 1896, 10.

41. "New Hippopotamus in Town," *New York Times*, December 5, 1896, 9.

42. "Baby Hippopotamus Takes to the Bottle: Removed from Tank to Mr. Smith's Office," *New York Times*, February 5, 1893, 10.

43. Minutes and Documents of the Board of Commissioners of the Department of Public Parks, 1987, 303, Parks Archives; "Fatima Goes to Chicago," *The Sun* (New York), April 8, 1897, 5.

44. "Hippo Sirus Made to Move," *The Sun* (New York), May 21, 1897, 7.

45. "Hippo Expects the Stork," *The Sun* (New York), May 2, 1899, 1; "Caging a Hippopotamus," *Freeman's Journal* (Dublin), July 10, 1899, 6.

46. "Baby Hippo's Trip to Germany," *The Sun* (New York), November 23, 1899, 7.

47. "Keeping the Family at the National Zoo," *Nature Magazine* 1 (1923), 25.

48. "Keeping the Family at the National Zoo."

49. As this paper traces the shifts in how hippo infant care was performed in zoos, it also demonstrates how separation of mother and infant occurred under inherently unnatural circumstances. Outside of captivity, infants nurse for nearly two years and stay with their mother for several additional years after that, but zoos often made decisions about separation based on the financial viability of offspring.

50. "New Baby Born at the Zoo," *Evening World* (New York), April 28, 1900, 3; "Fourth Baby Hippopotamus Born to Caliph and Miss Murphy," *St. Louis Dispatch*, May 26, 1900, 41.

51. "Baby Hippo in Central Park," *Buffalo Evening News*, May 4, 1900, 4.

52. "Big Park Hippos in a Fight," *The Sun* (New York), September 21, 1901, 1; "Hippos in Fierce Fight," *Farmer City (IL) Journal*, October 11, 1901, 7.

53. "They Also Play Tag: 'Jim' Crowley of New York's Central Park Menagerie and Caliph II Daily Entertain Children and Grown-Ups at the Metropolis," *Bucklin (KS) Banner*, September 4, 1913, 7.

54. "Collection of Central Park Zoo, New York City," 1914, Books and Manuscripts, New York City Department of Records and Information Services.

55. "New York's Biggest Baby Is Sold to Circus," *Knoxville (TN) Sentinel*, December 1, 1916, 4.

56. "How Animals Are Secured for Zoological Parks," *New York Times*, September 6, 1903, 28.

57. "A Hippo's Happy Family," *Kansas City (MO) Star*, May 5, 1905, 14.

58. "Sitting Up with Miss Murphy," *The Sun* (New York), 10, 1908, 3.

59. "Uniform Zookeeper (Male)," 1936, New York City Parks Department Photo Archives. This sex-segregated workspace was prevalent in zoos throughout the country. On the other side of the city, Helen Martini documented a similar experience at the Bronx Zoo. As the wife of the lion zookeeper, she also performed unpaid care work, often out of her home, for infant animals at the zoo. "Nursemaid to a Zoo," *St. Albans (VT) Daily Messenger*, January 22, 1947, 7.

60. Daniel Bender recounts a similar care work trajectory for women in his overview of zoos in the opening years of the Cold War. Daniel Bender, *The Animal Game: Searching for Wilderness at the American Zoo* (Cambridge, MA: Harvard University Press, 2016): 237–71.

61. "Lehmans Donate Children's Zoo," *Corsicana (TX) Daily Sun*, June 16, 1960, 5.

62. "Ex-Debs Animal Kingdom," *Oakland (CA) Tribune*, December 4, 1968, 71.

63. "Swapping Beasts Keeps Zoo Stocked," *New York Times*, February 7, 1915, 7.

64. Minutes and Documents of the Board of Commissioners of the Department of Public Parks, April 30. 1889, 9, Parks Archives.

65. "Central Park Animals Sold," *Brooklyn (NY) Daily Eagle*, June 27, 1902, 1.

66. "Bargains in Wild Animals," *Tyrone (PA) Daily Herold*, June 26, 1919, 1; "Zoo Lioness Fails to Raise a Bid," *Times Union* (New York), June 28, 1929, 3.

67. "Stork Brings Baby Hippo," *Topeka (KS) State Journal*, November 7, 1911, 14.

68. "Bill Snyder of Central Park Zoo to Quit," *The Sun* (New York), February 3,

1918, 61; "Swapping Beasts Keeps Zoo Stocked," *New York Times,* February 7, 1 915, 7.

69. "At What Time in Life Does an Animal Grow Old," *New York Times,* February 12, 1911, 10; "Two Hyenas Executed," *Brooklyn Daily Eagle,* January 30, 2911, 4.

70. William Conklin, "The Central Park Menagerie," *The Epoch,* June 29, 1888, 406.

71. "Zoo's Big Buffalo Sold to a Butcher," *New York Times,* November 10, 1915, 14.

72. "Vicious Elephant Killed," *Daily News* (Lebanon, PA), October 4, 1902, 1; "Elephant Not Spared," *Silver Blade,* July 9, 1915, 1; For a deeper look at the history of punishment and execution of male elephants in captivity, see Susan Nance, *Entertaining Elephants: Animal Agency and the Business of the American Circus* (Baltimore: Johns Hopkins University Press, 2013).

73. For a broader look at intersections of zoos, museums, and taxidermy, see Rachel Poliquin, *The Breathless Zoo: Taxidermy and the Cultures of Longing* (Philadelphia: University of Pennsylvania Press, 2012); Jay Kirk, *Kingdom under Glass: A Tale of Obsession, Adventure, and One Man's Quest to Preserve the World's Greatest Animals* (New York: Henry Holt, 2010); Karen A. Radar and Victoria E. M. Cain, *Life of Display: Revolutionizing U.S. Museums of Science and Natural History in the Twentieth Century* (Chicago: University of Chicago Press, 2014).

74. "Old Caliph's Skin a Museum Exhibit," *New York Times,* July 2, 1911, 4; "Giant Mounted Hippopotamus," *Nashville Tennessean,* March 27, 1910, A4.

75. "Mother of 16 to Be Stuffed," *Daily News* (Mount Carmel, PA), May 2, 1929, 5.

76. Milwaukee Public Museum Twenty-Eighth Annual Report, September 1, 1909–August 31 1910, 27, Milwaukee Public Museum Archive Collections, Milwaukee, Wisconsin.

77. Letter from Henry Ward to Al Ringling, February 26, 1910, Milwaukee Public Museum Archive Collections, Milwaukee, WI.

78. "Seeks Her 14 Children," *Times Union* (New York), April 4, 1928, 2.

79. "Baby Hippo Hits the Bottle," *Cushing (OK) Daily Citizen,* June 26, 1938, 8.

80. "This Hippo Can Snuggle," *Daily Tar Heel* (Chapel Hill, NC), October 21, 1948, 2.

81. "Whose Baby?" *Burlington (VT) Daily News,* September 17, 1953, 7.

82. "Central Park Zoo, Baby Hippopotamus," 1964, New York City Parks Department Photo Archives.

83. "Hippo Family Moving Out of Central Park Home to Make Way for Two

Pygmy Tenants," *New York Times*, July 30, 1974, 37; John C. Devlin, "Reluctant 10,350-lb Hippo Family Packed for Trip," *New York Times*, August 9, 1974, 37.

84. "Young Woman Has Career Caring for Primates," *Times Standard* (Eureka, CA), August 26, 1973, 10.

85. Bender, *The Animal Game,* 261.

86. See, for example, "River Hippopotamus (Hippopotamus amphibius) AZA Animal Program Viability Analysis Report," Association of Zoos and Aquariums, 2016.

87. David M. Powell and Matt Ardaiolo, "Tough Questions, Complex Answers: American Zookeeper Responses in a Nationwide Survey about Culling," in *Scientific Foundations of Zoos and Aquariums*, ed. Allison B. Kaufman, Meredith J. Bashaw, and Terry L. Maple (Cambridge: Cambridge University Press, 2019), 304–24.

88. Jesse Donahue and Erik Trump, *The Politics of Zoos: Exotic Animals and Their Protectors* (DeKalb: Northern Illinois University Press, 2006), 158–59.

89. Carrie Friese, *Cloning Wild Life: Zoos, Captivity, and the Future of Endangered Animals* (New York: New York University Press, 2013), 105.

90. Mads Frost Bertelesen, "Issues Surrounding Surplus Animals in Zoos," in *Fowler's Zoo and Wild Animal Medicine Current Therapy* 9 (St. Louis, MO: Elsevier, 2018): 134–36.

MARY TRACHSEL

seven The Destructive Ecology of Human-Pig Relations in Iowa since 1950

Science tells us that as a species, pigs (*Sus scrofa*) possess an array of cognitive abilities that overlap our own. Studies of porcine learning, memory, recognition of individuals, attention bias, cooperative problem-solving, temporal perception, open-ended categorization, deception, performance of joy-stick-operated video tasks, and responsiveness to human cues collectively reveal that pigs are comparable to or more sophisticated than many companion animal species in intelligence and social cognition.[1] A recent comparative review of what is known about pigs' psychological characteristics concludes that "pigs possess complex ethological traits similar, but not identical, to dogs and chimpanzees."[2] Yet pigs, for whom domestication equals commodification as food for human consumption, are rarely regarded as companions, like dogs, or as cognate personalities, like other species of great apes. Their exclusion from these preferred nonhuman categories means that pigs' experience, psychological as well as physical, has historically garnered little human attention. The current and ongoing spread of intensive hog-farming practices around the world, however, and with it the dramatic increase in the planet's pig population, have recently begun to focus some concern for nonhuman animal welfare on the industrial farming conditions in which most pigs live out their short lives today.

In the United States, by far the largest and most intensive hog farming occurs in my home state of Iowa, where the resident pig-to-human ratio is nearing eight to one. The growth of Iowa's hog herd is a fairly recent and rapid historical development, as pigs are not native to the North American prairie. They first appeared on the scene in the mid-nineteenth century, when small numbers of pigs accompanied the families of Euro-American settlers who initiated the transformation of prairie into farmland. Within a century, and largely through the interactions of pigs and people, this transformation was all but total. Although it is an agricultural state, Iowa might be considered the most "developed" state in the union, with only a miniscule fraction of its original prairie habitat remaining in small, widely dispersed and isolated scraps.

This essay traces the expansion of Iowa's domestic hog herd over the past century and a half, with special attention to its acceleration under industrial animal-farming methods from the latter part of the twentieth century to the present. As the state's most powerful economic driver, hog farming has steadily consolidated the production, processing, and marketing of pork into the hands of fewer and fewer people, who practice "vertical integration"—corporate control of the production process "from squeal to meal," as Smithfield Foods described it in a 2001 annual report.[3] Under this corporate model, the production of feed—primarily corn and soy—and the breeding, farrowing, nursing, and finishing of pigs for market take place on vast company-owned farms overseen by hired managers, while slaughter, processing, and packing also take place in company-owned facilities. In its geographical context, this industrial development poses an ever-growing environmental threat to human and nonhuman inhabitants of Iowa and beyond, while the economic and political might of the pork industry shields its inner workings from public scrutiny. Thanks to biosecurity regulations that keep factory hog farms closed to the public and "ag-gag" laws that criminalize the actions of those who seek to expose the living and dying conditions of these cognitively sophisticated animals in confinement operations and industrial slaughterhouses, most people in Iowa today have never met a pig in person. Outside of these carefully guarded production sites, most face-to-face encounters between pigs and people occur in curated, educational settings that are in many ways comparable to the zoos described elsewhere in this volume—in livestock shows at county and state fairs and in demonstration barns and displays that showcase the hygienic and high-tech modern production of pork.

HOW HOGS CAME TO LIVE WHERE THE BUFFALO ROAMED

The brief but intensely intertwined histories of pigs and people in Iowa exemplify the development of regionally important "mutually conditioning relations," an ecological term describing intra- and interspecies dynamics in geographical locations, or biomes.[4] Human and pig species first coinhabited Iowa in the second half of the nineteenth century, on prairie-sod farmland populated sparsely by people and even more sparsely by pigs. At the time of their joint arrival, the land had been inhabited by fourteen Indigenous plains and woodlands tribes whose seminomadic lives were tied to the movement of free-roaming grazers, notably deer, elk, and bison. At the time Euro-American colonization began, an estimated 85 percent of the Iowa landscape was tallgrass prairie. This diverse and complicated biotic community in an especially fertile portion of the US grassland was dominated by the interactions of tall grasses (e.g., big and little bluestem and switchgrass), grazing animals above ground (e.g., deer, buffalo and elk), and deep root systems and soil-dwelling creatures below. Today, less than 0.1 percent of this original habitat remains in the state, because tallgrass prairie has been heavily exploited as the richest cropland on the continent.[5]

Before the arrival of Euro-American settlers, Indigenous people "managed" the prairie by planting nonnative species such as squash and corn imported from Central and South America via extensive trade routes operated by the Cahokia community in the Mississippi flood plain near present-day St. Louis, Missouri. But except for the use of fire to maintain and expand the prairie eastward across the Mississippi, Native Americans' activities had little overall impact on the ecology of the prairie, as their gardens and villages were small, localized, and widely dispersed, and humans were vastly outnumbered by the large and small game animals they relied on for meat, fat, bones, feathers, and hides. Accordingly, as prairie historian Cornelia Mutel reports, before the effects of the Columbian arrival reached Iowa, "Native Americans were part of a sustainable, stable, and self-sufficient system, one where rivers ran clean, the prairie soils grew thicker and richer, and thousands of plants and animals continued their reproduction and massive annual migrations."[6]

European-American settlement first occurred in Iowa after the 1803 Louisiana Purchase from France of a wide swath of land west of the Mississippi River. In subsequent years, US government-sponsored military and germ warfare (i.e., smallpox), forced treaties, and removal of resident Native American tribes to "Indian territory" in Kansas opened the Iowa prairie to its eventual transformation by white settlers and European-style agriculture. Initially, Euro-American settlers, accustomed to woodland landscapes, were suspicious of the treeless plains, considering them barren and lacking timber for building homes and fences.[7] By the 1830s, however, this view had changed, as "farm-makers" from New England, the Central Atlantic states, and north-central Europe were discovering the fecundity of Iowa soil, built up from millennia of annually composted grasses and forbs and fertilized by a steady supply of rich manure from massive herds of buffalo, elk, and deer.

The transformation of Iowa's prairieland was swift and thorough, propelled in the mid-nineteenth century by the industrial manufacture of farm machinery to enable the deep-furrow plowing needed to disrupt the extensive root systems of native prairie plants. White settlers in Iowa soon aspired to replace the rugged wild prairie flora with grains and "tame grasses" such as wheat, barley, corn, and oats, a move replicated in the substitution of domesticated animals for the wild ones hunted by Indigenous people. To illustrate the speed of the American prairie's domestication, ecologist Candace Savage cites Meriwether Lewis's 1804 report from Iowa's neighbor, South Dakota, of encountering herds of the apex grazer, "bison without number," then notes that the last free bison in Canada was killed in 1883, and in the United States in 1891.[8] Less obvious to the human eye than the disappearance of buffalo and tall grasses was the cascading effect of this disruption throughout the prairie biome—the steady depopulation of smaller or less abundant species, including reptiles and amphibians, birds, mammals, insects, and the plants and protists that sustain their existence. Invisible to the human eye, the micro-biotic changes that accompany such rearrangement of the macro-biome are easily overlooked and often poorly understood.[9]

The agricultural replacement of wild with domesticated species was the beginning of a steady and continuous process of ecological simplification of the tallgrass prairie biome. On the preagricultural prairie, grazing was the process most responsible for maintaining ecological health and stability through bio-diversification. As Mutel explains, "Grazing created a patchwork of diverse environments: bare and vegetated soils, sunny and shaded sites, short and tall plant growth, and transitory and mature communities. The resulting mosaic of habitats fostered a diversity of species and maximized ecological stability. Without a variety of disturbances and habitats, native plant diversity and the land's stability and sustainability would have declined."[10]

The earliest stages of Iowa farming retained grazing as the land's central activity, replacing wild grazing animals with domestic grazers—notably sheep, beef and dairy cattle, and horses—that fed on native grasses. But the natural fertility of Iowa's soil soon favored the economic transformation of grazing to feed-based animal agriculture, and the iron plow and the tractor hastened the attendant transformation of flora from native grasses to imported grains.[11] As tractors began to replace horses throughout the latter half of the nineteenth and early twentieth centuries, the demand on pastureland gradually eased, allowing the expansion of herds of cattle and sheep or the repurposing of pasture acres as row-crop fields. While the nineteenth-century farmers of the tallgrass "prairie peninsula" (Iowa and portions of central and northern Illinois) were not monoculturists, regional specializations in crop and livestock production were gradually developing in the area.[12] In particular, farmers in the peninsula were discovering that pigs and corn were "strikingly complementary" and that the economic relationship between hog and corn prices made it "almost always profitable to feed hogs."[13] Iowa historian Allan G. Bogue goes on to explain that "in Iowa, one hundred pounds of pork was worth less than ten bushels of corn in only three years between 1861 and 1900."[14]

Initially, pigs comprised but a small portion of the livestock raised by Iowa farmers, with individual families fattening only a few pigs for their own consumption or local sale. Prior to 1830, hogs functioned primarily as bottom feeders or garbage disposals on family farms, so raising hogs was considered "an industry which is naturally carried on upon a small scale, as the hog is the means of utilizing and converting food matter and waste which would otherwise be lost."[15] As such, hogs began as a livestock animal secondary to cattle during the transformation of prairie first into pastureland and then into row-crop fields. But pigs' popularity advanced when German and Swedish farmers in particular became "competitively efficient" by fattening hogs on the food residue left by beef cattle, and Bogue concurs that raising pigs gave this ethnic farm population an economic edge that other farmers soon began to emulate: "What pulled most of the debt ridden out of the red was the combination of pluck, perspicacity and pigs."[16] In short, by excelling at waste disposal, expanding pig populations in states like Iowa eventually began to stimulate and satisfy human appetite for pork in a burgeoning capitalist society.

The steady growth of Iowa's hog population throughout the nineteenth and twentieth centuries and into the twenty-first has paralleled the expansion of its corn crop. Agricultural historian Earle Ross traces the origin of Iowa's rapid rise to national leadership in agriculture to "the particular adaptability of the region to corn growing," coupled with the discovery that the most profitable use of corn was for fattening hogs.[17] By the second half of the nineteenth century, Iowa had achieved the status of a "corn surplus" state within a larger midwestern area that was coming to be called "the corn belt."[18] From 1830 on, the herd size of hogs throughout the region grew steadily, prompting an 1834 observer to express wonder at the "immense droves" some farmers possessed in the Mississippi River Valley.[19] Eventually, the herds of some of the wealthiest farmers in the region increased to several hundred head. Even before a localized meatpacking industry began to expand across the state, spurring incremental growth in both the hog population and the corn crop, farmers found it profitable to translate corn into pork that could "walk to market," sparing them the cost of shipping and processing corn as human food or alcohol.

As pigs gained market value in the American Midwest, farmers devoted increased attention to their "improvement" as a food commodity. Throughout their history in the state, Iowa's pigs have been subject to a long process of selective breeding and genetic modification that would eventually result in severely reduced genetic diversity in the herd. This genetic engineering process began in the nineteenth century with selective breeding of stock from various parts of Europe and Asia. In 1891, John Shaffer, secretary of the Iowa State Agricultural Society, proclaimed that Iowa farmers had perfected pork production through selective breeding of hogs for fast and efficient growth, formal symmetry, and uniform size, all qualities that facilitated rapid, standardized pork processing.[20] Shaffer was speaking before the arrival of refrigeration, however, which reconfigured the market at the outset of the twentieth century and led to changes in the kinds of pork human consumers demanded. The shift from salt-cured to chilled meats produced a preference for hogs with less fat and more muscle,[21] and this preference coincided with the deterioration through overbreeding of the low-set, fine-boned "lard type" hog. After farmers experimented for a time with breeding large, heavy-boned, long-legged "bacon-type" and "meat-type" hogs, packers began to complain that the meat these animals produced was inferior and difficult to process, and they successfully lobbied breeders and agricultural colleges to cultivate lower finishing weights. To meet such demands, in 1936, Iowa State University established the United States Swine Breeding Laboratory in Ames, Iowa, with the mission of improving hogs on several measures of productivity and profitability, including increased litter size, increased piglet survival rate, faster growth rate, and more desirable meat quality.[22]

Such commodification-improvement measures paved the way for the factory model of hog farming that would eventually take hold in Iowa in the late twentieth century. Producing more pigs per breeding sow and moving those pigs more rapidly from farm to market to consumer's table meant that pork production was becoming a reliable path to wealth for hog farmers and pork processors and an affordable opportunity for human consumers to satisfy their appetite for meat. Of course, this evolution of the pork market brought changes to the lives of pigs themselves, as their life spans grew shorter, their living space dwindled and grew ever more crowded, and their opportunities for social and sensory stimulation eventually collapsed into the conditions of factory containment.

IOWA'S DWINDLING BIODIVERSITY: MORE OF LESS

Most economic historians trace the rise of industrial agriculture in Iowa to the 1970s, when a massive crop failure in the USSR created an export-market incentive for rapidly expanded production. To meet the international community's sudden demand for exported grain, Earl Butz, US secretary of agriculture from 1971 to 1976, famously urged farmers in the heartland to plant "from fence-row to fence-row," admonishing them to "get big or get out." Throughout the 1970s and 1980s, a period known to economists and rural Americans as "the farm crisis," many small farms were swallowed up by large farms on their way to getting bigger. The number of hog farms across the United States in the late 1970s was approximately 650,000, but by 2004, that number had dropped to below 70,000.[23] Iowa has participated in this national consolidation of agricultural land and production among fewer farmers, as the number of farms in the state declined dramatically throughout the latter half of the twentieth century. In 1945, the number of Iowa farms raising hogs (generally as part of a diversified animal farming operation) was close to 170,000; by 1997, that number had declined to a little over 17,000.[24] The trend continues to this day, as the Iowa Pork Producers' most recent estimate places the number of hog farms in Iowa at 5,400.[25]

Initially, despite devoting increased land and crops to pork production, the state was relatively slow to adopt the factory model of hog farming. Toward the end of the twentieth century, most of Iowa's pigs were still pasture-raised in A-frame huts and finished on concrete feed lots, but the state was poised to adopt the factory-farm model already taking hold in North Carolina. The Iowa legislature opened the door to factory farming from 1975 onward by steadily weakening laws prohibiting packers and processors from owning hogs or obtaining hogs from farmers on contract.[26] Once concentrated animal feeding operations (CAFOs) began to appear on Iowa's landscape, the disappearance of small farms accelerated. USDA statistics for Iowa indicate that the number of farms raising hogs in the state in 2012 was 6,616; within the next eight years, according to the figures of the Iowa Pork Producers Association (IPPA), that number declined by roughly 20 percent.[27] The bulk of Iowa pigs produced for market (69 percent) now comes from farms with an inventory of 5,000 or more. With a breeding sow herd of over 242,000, the state's largest producer, Iowa Select Farms, raises over four million hogs per year on 800 farms spread across fifty Iowa counties.[28]

The disappearance of the diversified family farm in Iowa and its replacement by factory farming operations is but the latest development in the unmaking of the tallgrass prairie's diverse ecosystem, anchoring the lives of human and nonhuman Iowa residents in a simplified food chain of corn and soybeans to pigs to people. As a result of this transformation, Iowa has become a site of environmental degradation that affects not only the state but the entire Mississippi and Missouri drainage system and the Gulf of Mexico. The nineteenth and twentieth centuries saw the installation of underground drainage tiles in 96 percent of Iowa's wetlands, a development that homogenized the land to make it "uniform in use and value."[29] Beyond reducing biodiversity, the loss of marshlands that functioned as wetland "sponges" has increased the threat of flooding in Iowa, while conversion of pastureland to row crops and the all-but-total eradication of deep tallgrass root systems has hastened the erosion of Iowa's topsoil, currently estimated at a rate of 5.2 tons per acre per year—1.3 tons more than the average among states in the Corn Belt.[30]

Meanwhile, expansion of the state's hog herd and the corn and soybean crops that feed them has brought a dramatic increase in agricultural pollution that has further diminished biodiversity in the state. Concentrated populations of pigs produce concentrated volumes of manure that overtax the environmental services Iowa's compromised waterways and drainage systems are able to provide. Fish kills have become frequent occurrences in Iowa's rivers and streams, many of them directly traceable to manure spills and leaks from large hog confinements' manure-management systems. A feeding operation housing five thousand pigs produces an amount of raw sewage equivalent to that of a town of twenty thousand, and the total manure production by Iowa's entire hog population has been estimated as the equivalent of waste produced by 84 million people.[31] Unlike Iowa towns, however, CAFOs are not required to run sewage processing plants. Much of the manure produced by Iowa pigs is applied as fertilizer to Iowa fields, but the system is challenged by problems of transportation, field saturation, spillage, and runoff. While the factory model maximizes hog-farming profits by breeding, feeding, and slaughtering pigs continuously throughout the year, their manure and other refuse—including urine, blood, and placenta—cannot be absorbed by Iowa's frozen fields in winter months; instead, it collects in liquid form in massive underground "lagoons" that are vulnerable to the spring flooding that is increasingly common and severe in Iowa. Beyond hog manure, other pollutants produced by industrial farming include "nutrients" such as nitrogen and phosphorous in commercial fertilizers; toxins for controlling weeds, insects, and parasites; and hormones and antibiotics to promote animal growth and boost pathogen resistance. All these substances combine to alter the chemical composition of Iowa's air, water, and soil, transforming many rural parts of the state into a hostile environment for people and the remaining birds, animals, plants, pollinators, and microbes that evolved on the tallgrass prairie.[32]

Karen Perry Stillerman, writing for the Union of Concerned Scientists (UCS), ranked Iowa high (fifteenth out of fifty) for federal investment in sustainable agriculture but noted that its pollution levels nevertheless remain among the country's worst. The UCS ranked Iowa forty-seventh in the nation for successful conservation practices and forty-nineth for successful reduction of ecosystem impact from human activity. An explanation for the discrepancy between investment and results, Stillerman writes, is the sheer scale of Iowa's agricultural industry. Indeed, industrial pig farming in Iowa has grown so large that its environmental pollution cannot be contained by the state or even the continent.[33] Bordered by the Missouri and Mississippi Rivers, Iowa is the main contributor to the nitrate load responsible for eutrophication, the process behind the formation of a dead zone in the Gulf of Mexico.[34]

Iowa's inability to contain the organic and synthetic waste produced by its intensive agro-economy is both a result and a reinforcement of the state's simplified ecology, as reduced diversity among domestic and wild life-forms leads to various forms of environmental degradation and dysfunction within the agro-ecosystem. Removing trees on the borders of fields, for instance, exposes fields to wind erosion, while reduction in the diversity of plant life renders crop-eating insects less subject to natural control. Such conditions prompt increased use of chemical fertilizers and insecticides, which enter the local air and water and disperse widely. The types of crops that dominate Iowa's agricultural landscape intensify the problem even further, as application of herbicides and synthetic fertilizer is higher for corn and soybeans than for any other crops in the United States.[35]

The loss of biological diversity due to agriculture in Iowa does not simply apply at the species level. In both hog and crop farming, industrial standardization of product means that biodiversity is lost at the genetic level too, a development that further weakens species' survival chances in a natural community. The genetic homogenization attending modern agribusiness thus raises the risk of zoonosis (nonhuman-to-human spread of microbial disease) beyond the level already imposed by dense livestock populations, as reduced genetic diversity favors increased viral potency. Laura Spinney notes that nearly all pigs raised in CAFOs today are white durocs, selectively bred to exhibit a few desirable physical and behavioral traits. "If a virus gets introduced to such a flock," she explains, "it can race through it without meeting any resistance in the form of genetic variants that prevent its spread."[36] This in turn can ratchet up microbial virulence, especially if the flock or herd is so large and dense that killing a host animal fails to check the virus's potency, other hosts being abundant and readily available.

The risks of pigs' genetic standardization are primarily borne by the pigs themselves. The industrial model's intensified efforts to standardize the size, shape, and growth rate of pigs, to maximize the size of their litters, and to adjust the quality of their meat have undeniably changed the lives of Iowa's pigs for the worse. Piglets bred for more efficient metabolism of commercial feed spend very little time with their mothers in order to free sows to produce between two and three litters per year. To further maximize efficiency, breeding sows are confined for nearly their entire lives for the sole purpose of gestating and feeding large litters of piglets. Iowa is among the forty-one states in the nation where gestation crates are legal. These are small cage-like enclosures that protect irritable pregnant sows in unnaturally close quarters from attacking one another, keeping them separate and largely immobilized, in part to reserve their nutritional intake for the production of piglets. Increased litter size has real effects on the welfare of piglets as well. The size of litters can compromise the fitness of individual piglets, as suckling piglets in large litters often fail to thrive and may even starve due to the inaccessibility of a functional teat. Mortality from sibling competition and crushing also increases with the size of the litter.[37]

These risks of intensive breeding have led to tighter control of pigs' lives—for example, their confinement in gestational and farrowing crates, and body modifications such as tail docking to prevent pigs from mutilating one another. Beyond these strictly controlled conditions, pigs' lifespans have been drastically shortened by evolving production processes. In 1923, a mast-fed razorback reportedly required two to three years to mature to a market weight of 200 pounds, while farm-fed pigs could attain the same weight in a mere eighteen months.[38] By comparison, most feeder pigs today reach a market weight of between 250 and 285 pounds when they are approximately six months old, thanks to selective breeding and genetic engineering, as well as feeding regimens for rapid weight gain.

AS ECOLOGICAL DIVERSITY DECLINES, HUMAN DIVERSITY RISES, AND IOWA FEEDS THE WORLD

While Iowa's millions of pigs are bred, raised, and transformed into human food as fast and efficiently as possible, processing operations follow a similar trajectory. In March 2019, Prestage Foods of Iowa described its new pork-processing facility in Eagle Grove as an example of "lean manufacturing" that requires only 920 workers to perform work that required 1,200 in a state-of-the-art facility twenty years prior. Using machinery such as "automatic openers, automatic neck droppers, robotic splitting saws, automatic rib pullers and water jet belly cutters," the new plant has reduced the "stun-to-cooler flow" of pig carcasses to a mere thirty-four minutes and has set a single-shift target of 10,000 hogs processed by 920 employees.[39]

The still-expanding Iowa pork industry's demand for disassembly-line workers in slaughterhouses and packing plants since the rise of CAFOs in the 1990s has diversified the human population of the state, even as its nonhuman diversity has diminished. A historically white and rural state since its settlement by farmers of northern European extraction, Iowa has experienced both urbanization and ethnic diversification in recent decades, both trends traceable to the growth of the state's hog-production industry. As family farms have consolidated into corporate entities and as the environmental quality of life declines due to agricultural pollution of air and water, many of Iowa's rural residents have relocated to the state's relatively small urban centers or have left the state altogether. The steady departure of people from rural areas can be measured in the ongoing consolidation of schools in Iowa's rural counties, along with the disappearance of hospitals, professional services, and local businesses. Throughout the state there is general anxiety about a "brain drain" from the loss of young, educated people who leave Iowa for better employment and living opportunities elsewhere, and this concern is especially focused on sparsely populated areas of the state. An anthropological study of my hometown in rural northeast Iowa describes these trends as "hollowing out the middle" of America.[40]

While the still-growing population of pigs across Iowa cannot singly explain these human population trends, the correlations are not accidental. Throughout the late nineteenth and early twentieth centuries, as the human population increased in settled rural areas, the number of hogs tended to decrease. Nearly a century later, an inverse correlation describes the geographical relationship of pigs and people in Iowa: as the hog population increases, the human population tends to decrease. Various reasons converge to explain people's flight from rural Iowa, including social isolation, limited access to essential services, and aesthetic and health concerns related to lax state regulation of water and air pollution, but as urban planning expert Charles Connerly sums it up, "Generally, people prefer not to live around pigs."[41]

The economic burden on humans who do live around large herds of pigs in Iowa has recently been assessed by the Union of Concerned Scientists (UCS) as the cost of meeting federal water quality standards. Iowa's rural human population is sparsely scattered over geographically large water districts and reliant on private wells and small-town water systems. Because of their proximity to large animal-feeding operations, these water sources are directly affected by agricultural runoff, but the heavy cost of cleanup and maintenance is spread among few residents. UCS economists calculate rural Iowans' annual per capita share of the projected costs of correcting EPA violations will be approximately six hundred times the average per capita share of urban Iowans ($1,200 as opposed to $2 per person per year).[42]

But somebody has to live near enough to pigs in Iowa to work at the CAFOs, slaughterhouses, and packing plants where Iowa pigs are transformed into enough pork to "feed the world." The process by which that transformation occurs is piece-by-piece disassembly of every pig that enters the stun-to-cooler flow. Job openings listed at the Prestage Foods Plant in Sioux City include butt boners, neckbone lifters, front-foot saw operators, pancreas pullers, and jowl shavers.[43] Increasingly the workers recruited to perform these jobs come from outside the country, primarily from Mexico and Guatemala, but also from Asia, Africa, and eastern Europe. As Connerly notes, "The state's agro-industrial economy depends on the availability of foreign labor," particularly as Iowa's human population ages and urbanizes.[44] In recent decades, the state's modest population growth has been almost entirely due to its foreign immigrants, many of whom are employed in meat industry work that is unskilled, physically difficult, dangerous, and poorly compensated. Of these foreign workers, a significant percentage are undocumented and may be paid less because of it, to offset the legal risk employers incur for hiring violations. Iowa has witnessed a number of packing plant raids by Immigration and Customs Enforcement (ICE), the most significant for the pork industry being the 2006 raid of the Swift pork-packing plant in Marshalltown, where ninety undocumented workers were arrested and deported.

Globally, human migration is a response to environmental pressures largely attributable to increased human population, consumption, and competition for control of resources, all compounded by the effects of climate change. Those migrants who find essential work in Iowa's hog industry necessarily live in close proximity to pigs, where they assume a disproportionate share of environmental risk, including the risks posed by air and water pollution and viral disease spread. At the same time, these workers are forced by social and economic vulnerability to endure workplace conditions that further threaten their well-being, even as their labor preserves and enhances the environmental, economic, and social well-being of industry owners and operators whose involvement takes place at a distance. A 2020 feature story on Iowa's hog barons noted that the owners of Iowa Select, the state's largest pork-production operation, live in a home overlooking a golf course in the suburbs of Des Moines and own several other homes in Naples, Florida.[45] Smithfield Farms, another large pork-production enterprise active in Iowa, is owned by the Hong Kong–based WH Group.

The dangers of pork-production work to humans who live near pigs are detailed in Ted Genoways's *The Chain*; they include repetitive motion injuries, exposure to deafening noise, loss of digits and limbs, and a neurological disorder known as PIN (progressive inflammatory neuropathy), an autoimmune-mediated response to "aerosolized pig neural protein" produced in the extraction of brains from carcasses.[46] To this list can now be added exposure to the COVID-19 virus. In April 2020, when a number of US packing plants in the Midwest were experiencing COVID outbreaks, one Iowa plant reported as many as 58 percent of workers testing positive.[47] Citizens in nearby communities called for closure of the plants, which were experiencing rising absenteeism due to illness or fear of infection. Some plants were temporarily shut down by public health authorities in April, and by early May, state news agencies were reporting that more than 1,600 workers at four Iowa meatpacking plants had been stricken by COVID-19, some of whom did not survive.

Like many other "essential workers" in the food supply chain, Iowa's pork-production line workers are disproportionately nonwhite, immigrant, and socially disadvantaged. In general, they are heavily dependent upon the wages they earn for essential labor that takes place at the physical interface of pigs and people, where they viscerally experience the mutually conditioning relationships of these species. Their frontline work increases their risk of exposure to interspecies bacterial (e.g., MRSA) and viral (e.g., swine flu) exchange, while the size, density, and mobility of the Iowa swine herd increases the risk of fast and extensive epizootic spread that could result in herd depopulation and economic shutdown.

Infectious disease outbreaks are not uncommon in today's intensive agriculture industry, but until COVID-19, they did not draw sustained public attention in the United States because they had not directly affected human lives. When mad cow disease, bird flu, or African swine fever sweeps through flocks or herds, the most effective eradication method is rapid shutdown of transport, along with wholesale onsite depopulation of animal inventories—that is, slaughtering and disposing of animals before they can be taken to market. As recently as the spring of 2019, to guard against the possibility of having to implement such a plan in North America's hog population, the US pork industry chose to cancel the International Pork Congress in Iowa's capital, Des Moines, fearing the introduction of African swine fever (ASF) through attendees and product samples from Europe and Asia, where the disease has already spread.[48] The Iowa Pork Producers' Association (IPPA) endorsed the industry decision, explaining to its members that depopulating Iowa's hog herd (then numbering approximately 23 million head) would have a "catastrophic" impact on the state's $36.7 billion pork economy, imperiling 142,000 jobs (more than 8 percent of the state's total) and affecting the state's enormous corn and soy sectors.

Though COVID-19 does not affect pigs, its initial spread through Iowa's human population did result in some depopulation of Iowa's swine herd because of bottlenecks in the supply chain. As pork-processing lines in plants slowed or stopped, producers had nowhere to send their market-weight hogs to make room for young feeder pigs ready to take their place in the CAFOs. In April 2020, these stoppages began to make real the grim prediction of the Iowa Pork Producers Association: "The specter of hogs being set loose or euthanized is a nightmare that will happen if there is not swift and meaningful intervention."[49] Though no releases of captive hogs have come to light, some farms were forced to euthanize portions of their herds, resulting in projected industry losses of $2.1 billion. On April 28, 2020, federal intervention finally arrived in the form of Executive Order 13917, mandating continued operation of meat- and poultry-processing facilities, which were officially designated as part of the critical infrastructure guaranteeing a safe and sufficient food supply to the American people.[50]

In reality, the market for Iowa pork reaches far beyond the American pork consumer, and this has never been truer than now. In an informational video on its "We feed the world" marketing slogan just weeks before the presence of COVID was declared in the United States, the industry summarized the USDA's report of a record $7.1 billion in pork exports, including a 149.9-million-pound increase in sales to China. The spike in Chinese demand for US pork stems directly from China's depopulation of its domestic herds, beginning in 2018, to stop the spread of African swine fever. In competition with European and South American producers, the US pork industry has been eager to seize the market moment. In 2019, pork and pork-product exports rose by 9 percent to record sales, with the Chinese market accounting for much of the increase, having more than doubled its demand for imported pork.[51]

The arrival of COVID in America's agricultural heartland did little to stop this expansion. Neil Dierks, CEO of the National Pork Producers Council, endorsed the "essential workers" classification established by Executive Order 13917, not just as a way to maintain the national supply chain but also as a way to continue growing the industry: "We recognize worker safety is paramount, but it's imperative that we maintain and increase harvest capacity for hogs."[52] Industry growth proceeded relatively uninterrupted. In tandem with claims that meatpacking line workers were "essential" because they ensured local and national availability of meat, the USDA reported in September 2020 that "U.S. pork exports to China hit an all-time record in just the first five months of 2020."[53]

The coincidence of human (COVID) and hog (ASF) pandemics in Iowa's global economic landscape was in many ways predictable. Just as close, prolonged, indoor contact intensifies the spread of viral disease among humans, many properties of contemporary hog farming in the United States, China, and parts of Europe and South America increase the probability of a "transmission event" in which an infectious disease escapes from one herd and spreads to others. These characteristics include industry expansion and intensification through increasingly concentrated pig populations and the movement of pigs through facilities that specialize in discrete stages of their growth. These conditions have steadily intensified over the past forty years, as the typical herd size for hogs in the United States increased from around two hundred in 1980 to over three thousand in 2017, a shift accompanied by eleven times more hog transport in 2015 than in 1980.[54] Herd expansion has been facilitated by the rapid consolidation of slaughter and packing activities in ever larger plants capable of processing bigger and faster streams of pork products to domestic and foreign markets. Between 1977 and 1997, the portion of America's hogs slaughtered in plants capable of processing more than one million animals per year grew from 38 percent to 88 percent. In 2020, five of the fifteen biggest slaughtering and packing operations in the country were in Iowa, and these five plants combined had a slaughtering capacity of 98,400 pigs per day.[55]

In classifying slaughterhouse and packing-plant workers with other essential, frontline food-industry workers, the pork industry has publicly deemphasized its international expansion and emphasized its contribution to the domestic food supply. In Iowa, the IPPA "Statement on Packer Operations Support," for instance, cited the complexity of the nation's food supply chain and the urgent need to maintain it, "especially those portions that operate in Iowa to provide food to Americans."[56] As COVID settled into the state's human population, a press release posted on the IPPA web page explained how American pork consumers would personally be affected if the slowdowns and stoppages were to continue: "With pork processing down by nearly 40 percent, we are in an unprecedented situation where farmers cannot get pigs processed so pork cannot get to grocery stores and to our end customer, you."[57]

NUTRIENT-RICH, ECOLOGICALLY IMPOVERISHED

Iowa, at the heart of the nation's heartland, is the biggest player in the pork industry's plans for growth. In its 2010 global report on animal agriculture, *Livestock's Long Shadow,* the United Nations' Food and Agriculture Organization (FAO) described fertile geographical locations like Iowa as "nutrient overload" sites, where the concentrated production of livestock creates an "unfavorable ratio of land to livestock." The ratio is "unfavorable" because the agricultural waste produced exceeds the land's capacity to absorb and use its "nutrients"—especially phosphorus and nitrogen. The FAO report maintains that the current trajectory of the global meat market is environmentally unsustainable, given the growth of the world's human population and the toll animal agriculture already takes on ecosystems. It is thus imperative, according to the FAO, that people reduce their meat consumption and begin to rely more heavily on a plant-based diet.[58]

Iowa's "nutrient reduction strategy" has so far failed to address the problem detailed by the FAO. Connerly has described this strategy as "a well-researched goal with no accompanying plan of action," adding that the authors of the plan "appear to expect that science will provide the solution to nutrient overload" with "new ideas that are cheaper and more effective" than those already projected.[59] With this unsubstantiated faith in technological solutions that obviate the need for human behavioral adaptation to environmental conditions, Iowa's pork producers aspire to grow their market even more. A strategy behind this market expansion is to define meat as an essential component of a wholesome diet, thereby rejecting plant-based eating as a threat to public health.

To be sure, consumption of pork is the primary physical interface in Iowa's human-hog relationship story, as pork producers attempt to boost consumption with the message that pork is not just good, it's good for you. This is a message Iowa senator Joni Ernst underscored in her proposed addition to the 2016 military spending bill, which required the US Defense Department to offer animal protein options in military facilities seven days per week. Ernst's proposal responded to the Coast Guard's adoption of "Meatless Monday," a public health initiative promoted by Johns Hopkins, Columbia, and Syracuse Universities in accordance with the UN endorsement of shifting to a plant-based diet. Seeking a preemptive ban on policies that promote reduced meat consumption by employees of government agencies, Ernst framed the consumption of meat as something US soldiers "deserve" and "need" to stay fit.[60] In an email explaining her proposal, she wrote, "The push for 'Meatless Mondays' in our military is misguided at best, and goes against military guidelines. Our men and women in uniform should have the option to consume the protein they need, including meat on a daily basis."[61]

In reporting the story to Iowans, one journalist noted that Ernst's proposal came at a time when obesity among active-duty military was at an all-time high, having risen 61 percent between 2002 and 2011, while 25 percent of young adults between the ages of seventeen and twenty-four were too heavy to meet the US military's physical requirement. Current CDC figures indicate that the situation has not improved since 2016, with 20.6 percent of children ages twelve to nineteen now qualifying as obese, while the percentage of obese twenty- to thirty-nine-year-olds is 40 percent (CDC). Of course, meat alone is not responsible for the nation's obesity problem, but a clear connection has been shown between the availability of inexpensive food that is high in saturated fats and Western countries' rising obesity rates, attended by an elevated incidence of chronic diseases such as type 2 diabetes, heart disease, elevated blood pressure, stroke, and some cancers, particularly colon cancer.[62]

PIG TO PLATE ADVENTURES

Public awareness of the health risks associated with meat-heavy diets, ecological understanding of agriculture's environmental impact, and human sensitivity to nonhuman sentience all pose public relations challenges to the pork industry today. One industry response to such challenges is the National Pork Board's "Operation Main Street," a campaign to grow consumer demand for wholesome pork while countering the voices of misinformed "activists." The program advances the commodification narrative of human-hog relationships by emphasizing the delicious outcomes of pork production and by tying the pork-production process to the hard, honest work of community-minded farm families who are well informed about environmental and agricultural science and technology. Many Operation Main Street events feature pork-centered meals—with local industry leaders grilling brats or handing out ham sandwiches at community picnics, or distributing ham dinners to regional food pantries during the holidays.

Often the food service at such events accompanies an educational program. A popular method of delivery is for farm families hosting a public feed to tell a story of personal growth in the industry and lead a guided tour of the farm, sometimes offering guests a rare chance to meet pigs face-to-face. Despite Iowa's millions of pigs, these designed encounters are a first-time opportunity for many Iowa consumers. Fear of disease spread through industrial-size CAFOs merits strict biosecurity measures that keep visitors at bay. These regulations are reinforced by legislation in Iowa and other agricultural states that classifies trespassing on agricultural facilities as domestic terrorism putting the nation's food supply at risk. Familiar to Iowans as "ag-gag" laws, such legislation criminalizes acts of photo documentation and surveillance by activists who spread bad publicity that threatens farmers' livelihood.[63] Animal rights and welfare advocates complain that as a result of such restrictions, Iowa's interstates are the best place for live hog sightings in the state. There, a report explains, drivers who pull abreast of livestock-hauling big rigs can see pigs "crammed into transport carriers, their noses sniffing at the metal ventilation holes for what will likely be their first and last breath of fresh air."[64] As animal activism has recently emerged as the "number one challenge to the future of the pork industry," industry leaders advise farmers to protect themselves by "YouTube-proofing" their property, increasing farm security, and consulting with local law enforcement authorities to create a crisis plan. Finally, they recommend "putting a face on animal agriculture."[65]

In Iowa, efforts to personalize corporate hog farming began with the arrival of CAFOs. In 1991, the Iowa pork industry in partnership with Kirkwood Community College staged the first farrowing display in the Swine Barn at the Iowa State Fair, now a popular annual event where fairgoers can witness the birth of piglets to a succession of sows as "one segment" of the pork production process. The farrowing segment begins when the inevitable slaughter of the unweaned piglets is approximately six months in the future, too distant for contemplation in the presence of adorable infants. The consumer education director for IPPA explains that the display is most visitors' first encounter with animal agriculture "in a real-life setting," and as such it presents an unusual opportunity for the public to authentically witness modern pork production.[66] In truth, the birthing place of these display litters is more spacious than "real-life" factory-farming conditions allow, and the sows in the display are granted a measure of privacy not available to those in CAFOs, who are hemmed in by their farrowing barn mates.

Since the farrowing display's inception, similar consumer education displays and visiting programs have proliferated in Iowa. These are the products of Operation Main Street's program to train pork producers and veterinarians "to tell their story . . . to improve the image of pig farming."[67] One such initiative is the annual Pig to Plate Adventure sponsored by the IPPA to introduce high school students to careers in pork production. Advertised activities include a real or virtual tour of a pig farm, talks with farmers and large-animal veterinarians, a pork fabrication demonstration (separating a pig carcass into meat cuts), hands-on pork cookery, and a pork recipe challenge (a recap of the 2020 online version of the Pig to Plate Adventure announces the winning recipes as "bacon-wrapped cheesy ranch potato stuffed pork loin," followed by "candy bacon empanada with maple-bacon ice cream"). While this public relations event is more specifically targeted than, say, a YouTube documentary in which a pork industry spokesperson interviews an Iowa farmer and his family while touring the farm and visiting young feeder pigs, both frame the pigs-and-people narrative as a decontextualized commodification story in which caring, community-minded professional farmers deliver high-quality protein to hungry humans.

Generally absent from such stories is an account of the environmental history and current state of the rural setting where the story unfolds, just as the exotic appeal of personal encounters with piglets obscures the day-to-day social deprivations and distortions of pigs' brief, commodified lives. To date, the economically powerful narrative of pigs as a food commodity continues to flourish in Iowa, unimpeded by environmental or animal-welfare activism. Citizen efforts to impose a moratorium on the building of new CAFOs in the state have so far proved unsuccessful, and there is little indication that the broader, escalating environmental concerns of the Anthropocene—climate change, species extinction, habitat destruction, pollution, and loss of natural ecological services—will outweigh the economic concerns of this Iowa industry anytime soon. Yet to be known are the possible long-term effects of vegetarian and vegan dietary movements, which have slowly but steadily expanded in the United States and elsewhere since Frances Moore Lappé's publication of *Diet for a Small Planet* in 1971. Reinforced by ethical concerns about animal welfare and the natural environment, could increasingly convincing, plant-based and cultured "meats" and the spread of meatless cuisine through cookbooks, cooking shows, and restaurant menus pose a serious economic threat to the pork industry? Iowa's story seems to suggest that only such a change in human diet could significantly alter the quality of life for pigs, as animals domesticated solely as a source of food.

NOTES

1. Ariane Veit et al., "Object Movement Reinactment in Free-Ranging Kune Kune Piglets," *Animal Behaviour* 132 (2017): 49–59; Elise Titia Gieling et al., "Assessing Learning and Memory in Pigs," *Animal Cognition* 14 (2011): 151–73; Sophie Brajon et al., "The Perception of Humans by Piglets: Recognition of Familiar Handlers and Generalisation to Unfamiliar Humans," *Animal Cognition* 18, no. 6 (2015): 1299–1316; Lu Luo et al., "Effects of Early and Later Life Environmental Enrichment and Personality on Attention Bias in Pigs (*Sus scrofa domesticus*)," *Animal Cognition* 22 (2019): 959–72; Melanie Koglmueller et al., "Are Free-Ranging Kune Kune Pigs (*Sus scrofa domesticus*) Able to Solve a Cooperative Task?," *Applied Animal Behaviour Science* 240 (2021): 105340; Natascha Fuhrer and Lorenz Gygax, "From Minutes to Days—the Ability of Sows (*Sus scrofa*) to Estimate Time Intervals," *Behavioural Processes* 142 (2017): 146–55; Marianne Wondrak et al., "Pigs (*Sus scrofa domesticus*) Categorize Pictures of Human Heads," *Applied Animal Behaviour Science* 205 (2018): 19–27; Suzanne Held et al., "Foraging Pigs Alter Their Behaviour in Response to Exploitation," *Animal Behaviour* 64 (2002): 157–66; Candace Coney and Sarah T. Boysen, "Acquisition of a Joystick-Operated Video Task by Pigs (*Sus scrofa*)," *Frontiers in Psychology*, February 20, 2021, https://www.frontiersin.org/articles/10.3389/fpsyg.2021.631755/full; Christian Nawroth et al., "Are Juvenile Domestic Pigs (*Sus scrofa domesticus*) Sensitive to the Attentive States of Humans? The Impact of Impulsivity or Choice Behaviour," *Behavioural Processes* (2013): 53–58.

2. Lori Marino and Candace Colvin, "Thinking Pigs: A Comparative Review of Cognition, Emotion and Personality in *Sus scrofa*," *International Journal of Comparative Psychology* 28, no. 1 (2015): 28.

3. Jeffrey J. Reimer, "Vertical Integration in the Pork Industry," *American Journal of Agricultural Economics* 88, no. 1 (2006): 234.

4. David Bello and C. Michelle Thompson, "Swarms, Herds and Peoples:

Examination of Interspecies Dynamics in China," *East Asian Science, Technology and Medicine* 48 (2018): 9–16.

5. Candace Savage, *Prairie: A Natural History of the Heart of North America,* rev. ed. (Vancouver: Greystone, 2020), 25.

6. Cornelia Mutel, *The Emerald Horizon: The History of Nature in Iowa* (Iowa City: University of Iowa Press, 2008), 11.

7. Allan G. Bogue, "Farming in the Prairie Peninsula, 1830–1890," *Journal of Economic History* 23, no. 1 (March 1963): 64.

8. Savage, *Prairie*, 13–14.

9. Savage, *Prairie*, 15–20.

10. Mutel, *Emerald Horizon*, 39.

11. Bogue, "Farming in the Prairie Peninsula," 62.

12. Bogue, "Farming in the Prairie Peninsula," 69.

13. Bogue, "Farming in the Prairie Peninsula," 69.

14. Bogue, "Farming in the Prairie Peninsula," 70.

15. Rudolf Alexander Clemen, *The American Livestock and Meat Industry* (New York: Ronald Press, 1923), 151–52.

16. Charles Wayland Towne and Edward Norris Wentworth, *Pigs: From Cave to Cornbelt* (Norman: University of Oklahoma Press, 1950), 211; Bogue, "Farming in the Prairie Peninsula," 71.

17. Earle Ross, *Iowa Agriculture: An Historical Survey* (Iowa City: Iowa State Historical Society, 1951), 75.

18. Clemen, *American Livestock*, 207.

19. Clemen, *American Livestock*, 53.

20. Ross, *Iowa Agriculture*, 75.

21. Towne and Wentworth, *Pigs*, 252.

22. Towne and Wentworth, *Pigs*, 258.

23. Brett Mizelle, *Pig* (London: Reaktion Books, 2011), 77.

24. Leana E. Stormont, "Overview of Hog Farming in Iowa" (East Lansing: Michigan State University College of Law, 1998), https://www.animallaw.info/article/overview-hog-farming-iowa.

25. IPPA, "2020 Iowa Pork Industry Facts," https://www.iowapork.org/news-from-the-iowa-pork-producers-association/iowa-pork-facts/.

26. E. Paul Durrenberger and Kendall M. Thu, "The Expansion of Large-Scale Hog Farming in Iowa: The Applicability of Goldschmidt's Findings Fifty Years Later," *Human Organization* 55, no. 4 (1996): 410.

27. USDA, *Census of Agriculture for Iowa, 2012*, https://agcensus.library.cornell.edu/wp-content/uploads/2012-Iowa-iav1-1.pdf; IPPA, *2020 Iowa Pork Economic*

Contribution Study, https://www.iowapork.org/wp-content/uploads/2020/09/2020PorkEconStudy_State_of_IA_Final_pdf.

28. Charlie Mitchell and Austin Frerick, "The Hog Barons," *Vox*, April 19, 2020; Iowa Select Farms, "A Commitment to Iowa and Our Rural Communities," https://www.iowaselect.com/webres/File/2020-a-commitment-to-Iowa-.pdf.

29. Charles Connerly, *Green, Fair, and Prosperous: Paths to a Sustainable Iowa* (Iowa City: University of Iowa Press, 2020), 42.

30. Connerly, *Green, Fair, and Prosperous*, 55.

31. David N. Cassuto, *The CAFO Hothouse: Climate Change, Industrial Agriculture and the Law* (Ann Arbor, MI: Animals and Society Institute, 2010).

32. Matthew Z. Liebman et al., "Using Biodiversity to Link Agricultural Productivity with Environmental Quality: Results from Three Field Experiments in Iowa," *Renewable Agriculture and Food Systems* 28, no. 2 (2013): 115–16.

33. Karen Perry Stillerman, *What Our 50-State Scorecard Says about Farming and Water Pollution* (blog), Union of Concerned Scientists, 2020, https://blog.ucsusa.org/karen-perry-stillerman/what-our-50-state-scorecard-says-about-farming-and-what-the-farm-bill-should-do-about-it.

34. Christopher Jones et al., "Iowa Stream Nitrate and the Gulf of Mexico," *PLOS ONE* 13, no. 4 (2018).

35. Liebman et al., "Using Biodiversity," 124.

36. Laura Spinney, "Is Factory Farming to Blame for the Coronavirus?" *The Guardian*, March 28, 2020.

37. Marco Ocepek et al., "Trade-Offs between Litter Size and Offspring Fitness in Domestic Pigs Subjected to Different Selection Pressures," *Applied Animal Behaviour Science* 193 (April 1, 2017): 7.

38. Clemen, *American Livestock*, 55–56.

39. Sharon Spielman, "How Prestage Foods of Iowa Built a State-of-the-Art Pork-Processing Plant," *Food Engineering.com*, March 4, 2019, https://www.foodengineeringmag.com/articles/98612-prestage-farms-built-a-state-of-the-art-pork-processing-plant-in-northwest-iowa.

40. Patrick J. Carr and Maria J. Kefalas, *Hollowing Out the Middle: The Rural Brain Drain and What It Means for America* (Boston: Beacon Press, 2009).

41. Connerly, *Green, Fair, and Prosperous*, 149.

42. Union of Concerned Scientists, "Rural Iowans Bear Brunt of Water Treatment Costs for Nitrate Pollution from Farms and CAFOs," press release, January 14, 2021, www.ucsusa.org/about/news/rural-iowans-bear-brunt-of-water-treatment-costs-for-nitrate-pollution-from-farms/.

43. Prestage Foods of Iowa, *Careers*, https://phg.tbe.taleo.net/phg04/ats/careers/v2/searchResults? org=PRESFARM&cws=37, accessed April 6, 2021.

44. Connerly, *Green, Fair, and Prosperous*, 128.

45. Mitchell and Frerick, "Hog Barons."

46. Ted Genoways, *The Chain* (New York: Harper, 2014).

47. Tony Leys, "Coronavirus Infects More than 1600 Workers at Four Iowa Meatpacking Plants," *Des Moines Register*, May 5, 2020.

48. US Department of Agriculture, *African Swine Fever (ASF)*, https://www.aphis.usda.gov/aphis/ourfocus/animalhealth/animal-disease-information/swine-disease-information/african-swine-fever/. "ASF is a highly contagious and deadly viral disease affecting swine herds around the world. The virus has spread from Sub-Saharan Africa to parts of Asia and Europe. It does not transfer to humans and is not considered a threat to food safety."

49. IPPA, "Covid-19 Pandemic Realities for Iowa Pork Producers," April 24, 2020, https://www.iowapork.org/we-content/uploads/2020/04/IPPA-COVID19-Sheet-updated-4.24.20.pdf.

50. Executive Office of the President, "Delegating Authority under the Defense Production Act with Respect to Food Supply Chain Resources During the National Emergency Caused by the Outbreak of COVID-19," *Federal Register*, May 1, 2020, https://www.federalregister.gov/documents/2020/05/01/2020-09536/delegating-authority-under-the-defense-production-act-with-respect-to-food-supply-chain-resources.

51. USDA, Foreign Agricultural Service, "China, Evolving Demand of the World's Largest Agricultural Import Market," September 29, 2020, https://www.fas.usda.gov/data/china-evolving-demand-of-the-worlds-largest-agricultural-import-market.

52. IPPA, "IPPA Statement on Packer Operations Support," April 17, 2020, https://www.iowapork.org/ippa-statement-on-packer-operations-support/.

53. USDA, Foreign Agricultural Service, "USTR & USDA Release Report on Agricultural Trade between the United States and China," October, 23, 2020, https://www.fas.usda.gov/newsroom/ustr-and-usda-release-report-on-agricultural-trade-between-the-united-states-and-china.

54. Gregory Cima, "Herd Sizes, Trade Risk Pig Health," *American Veterinary Medical Association Journal* 250, no. 9 (May 2017): 260.

55. Jayson Lusk, "These 15 Plants Slaughter 59% of all Hogs in the US," *Jayson Lusk* (blog), April 19, 2020, http://jaysonlusk.com/blog/2020/4/9/these-15-plants-slaughter-59-of-all-hogs-in-the-us, accessed May 15, 2021; Karen McMahon,

"Four Packers Kill 57% of Hogs," *National Hog Farmer,* November 4, 2021, http://www.nationalhogfarmer.com/farming_four_packers_kill.

56. IPPA, "IPPA Statement on Packer Operations Support," April 17, 2020.

57. IPPA, "How COVID-19 Has Impacted our Food Supply Chain and Iowa Pork Producers," May 8, 2020, https://www.iowapork.org/how-covid-19-has-impacted-our-food-supply-chain/.

58. United Nations, Food and Agriculture Organization, *Livestock's Long Shadow: Environmental Issues and Options*, 2010, http://www.fao.org/3/a0701e/a0701e.pdf.

59. Connerly, *Green, Fair, and Prosperous*, 74.

60. Erich Wagner, "Senator Seeks to Ban Nonexistent 'Meatless Monday' Policy at Agencies." *Government Executive*, April 16, 2021.

61. Christopher Doering, "Meatless Monday for Military? No Way, Ernst Says," *Des Moines Register*, June 13, 2016.

62. Vaclav Smil, "Eating Meat: Evolution, Patterns, and Consequences," *Population and Development Review* 28, no. 4 (December 2002): 626.

63. Ann Hess, "Pork Producers, Beware," *National Hog Farmer*, February 14, 2020, https://www.desmoinesregister.com/story/news/politics/2021/04/12/iowa-legislature-passes-new-ag-gag-agriculture-trespass-bill-governor-kim-reynolds/7099951002/.

64. Leana E. Stormont, "Overview of Hog Farming in Iowa," in David Favre and Rebecca Walsh, eds., *Animal Law* (East Lansing: Michigan State University College of Law, 2003), https://www.animallaw.info/article/overview-hog-farming-iowa.

65. Jennifer Shike, "Four Ways to Protect Your Farm from Activists," *Farm Journal: Pork Business*, February 6, 2020, https://www.porkbusiness.com/news/hog-production/4-ways-you-can-protect-your-farm-activists.

66. Shike, "Four Ways."

67. Vincent ter Beek, "Farm Visit: Opening New Avenues for Pig Promotion," *Pig Progress,* March 2021, https://www.pigprogress.net/World-of-Pigs/Articles/2021/3/opening-new-avenues-for-pig-promotion-7/8465E/.

JOHN M. KINDER

eight "The Next Meal for the Lions" The US Occupation of the Baghdad Zoo, 2003–2004

ZOO CRIME I: KILLING A TIGER

Early on the morning of September 19, 2003, Captain William (Wes) Sumner of the 354th Civil Affairs Brigade, the military's chief liaison at the Baghdad Zoo, arrived to find a state of panic. "Why did you do this? Why did you do this?" shouted the guard at the entrance, speaking less to Sumner than to all the men and women who wore the captain's uniform. Not long after entering the zoo, Sumner spotted a blood trail—first a few spots, then dark blotches where blood had pooled and dried—leading toward a maintenance entrance of the tiger cage. At the center of the dusty enclosure was the corpse of a male Bengal tiger, already stiffening in the morning heat.[1]

Scenes of bloodshed were common enough in Baghdad in 2003. Since the start of the US-led occupation that March, the Iraqi capital had suffered deadly attacks on a regular basis. Located within the Green Zone, the heavily fortified center of US operations in the city, the Baghdad Zoo had escaped the worst of the violence: the car bombings, the improvised explosive devices, the lethal tactics of Americans on patrol. Still, in a country riven by insurgency and civil war, nowhere was completely safe, not even behind the iron bars of a zoo cage.

Gathering the forensic evidence and interviewing staff, Sumner began to piece together what had happened: there had been a barbecue for US military at the zoo's island café. At some point in the evening, thirty-two-year-old Sgt. Keith Mitchell (possibly drunk, though he claims to have had only one beer) wandered over to the tiger cage with some of his buddies. According to the

Wall Street Journal, Mitchell maintained that he was innocently standing a few feet from the enclosure when, out of nowhere, his arm was violently pulled into the metal cage.[2]

Sumner's own analysis told a different story, however. Perhaps out of malice, perhaps out of boredom, most likely fueled by arrogance and stupidity, Mitchell had entered the "dead zone"—the space between the enclosure's inner and outer cages—and slipped his hand through the bars. That was when the tiger grabbed him, breaking his "arm above the elbow" and tearing off his middle finger. As Mitchell cried for help, one of his fellow soldiers shot the tiger through the left shoulder with an unauthorized pistol confiscated from an Iraqi. The bullet ripped a path through the tiger's chest cavity, and he died of internal bleeding. Sumner and the zoo staff performed an autopsy, located the pistol's slug, and buried the tiger shortly thereafter (fig. 8.1). The tiger was later exhumed on rumors that he had "ingested human remains," though no evidence was found (fig. 8.2).

The killing of the Bengal tiger was a major blow to the Baghdad Zoo's already meager collection of "charismatic megafauna," a term that conservationists often apply to large exotic animals with broad public appeal (e.g., elephants, tigers, lions).[3] The incident also had dire political implications. As Sumner later recounted, the soldier's lapse in judgment threatened to "undo months of hard work that [had] been invested in saving" Baghdad's zoo animals.[4] Most ominous of all, the tiger killing exposed the contradiction at the heart of American militarism: that the most heavily armed nation in world history fancied itself a force of liberation and global peace. If the Americans were willing to shoot an innocent tiger (hungry and helpless and caged), were they any better than the forces of Iraq's recently ousted dictator, Saddam Hussein?[5]

This chapter offers a critical history of the US occupation of the Baghdad Zoo in 2003–4. More than fifteen years on, the rebuilding of the Baghdad Zoo remains one of the few public relations highlights to emerge from the United States's calamitous invasion of Iraq. Even as American armaments reduced large sections of the Iraqi capital to rubble, a small contingent of American soldiers, former staff, foreign aid workers, and local volunteers risked their lives to save the zoo's thirty-five or so remaining animals. Throughout the spring and summer of 2003, US occupation officials trumpeted the zoo's reconstruction as evidence of America's civilizing mission in the region. Behind the scenes, however, Baghdad's zoo animals faced disease, the lingering effects of malnu-

FIG. 8.1. This photograph, one of more than sixty taken at the site of the tiger killing, reveals a side of the US occupation of the Baghdad Zoo that few would ever see. Taken a few hours after the event, it shows the efforts of zoo staff to remove the bullet from the tiger's chest. Image courtesy of William Sumner.

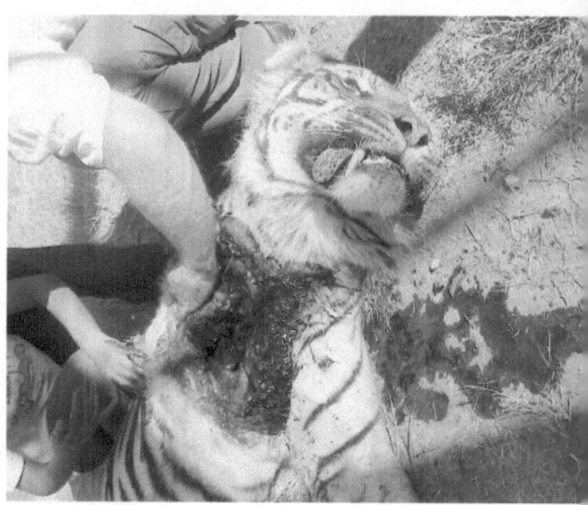

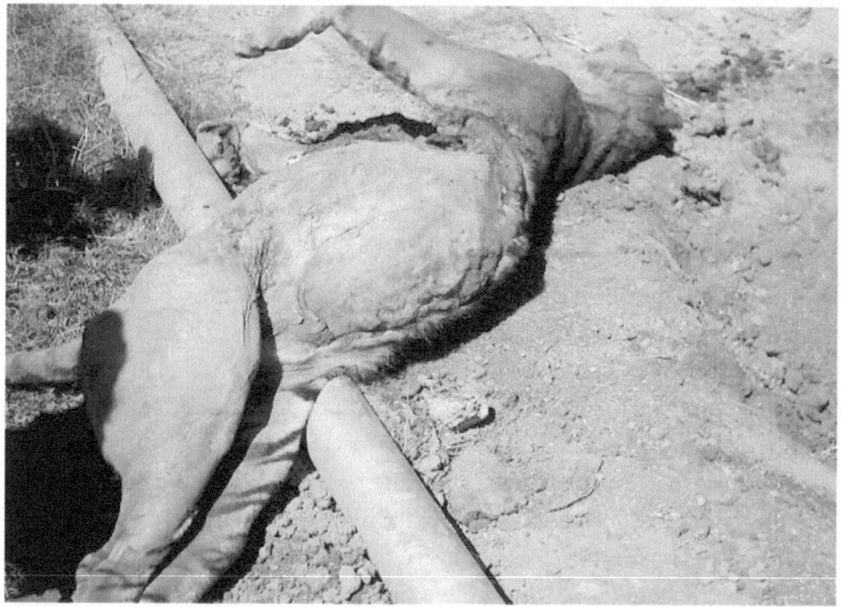

FIG. 8.2. This photograph shows the tiger's corpse after he was exhumed a few days after the initial autopsy. Zoo staff wanted to check whether the tiger had swallowed any of Keith Mitchell's finger (he hadn't). In the days following the episode, William Sumner believed that the tiger killing represented such a stain on America's image that he recommended taking immediate steps to "counter the potential for bad publicity before it takes on international significance." Image from Department of the Army, "Death of a Tiger at Baghdad Zoo," courtesy of William Sumner.

trition, and the routine threat of violent attack (from anti-American insurgents and from US forces themselves). Indeed, the hoopla surrounding the Baghdad Zoo belied growing tensions among Iraqis, animal rights organizations, and the image-conscious Coalition Provisional Authority (CPA) about the fate of the zoo's caged inhabitants.[6]

Much of my analysis relies upon a cache of government documents, military memos, and previously uncirculated photographs that I obtained while carrying out a series of interviews with major participants in the zoo-rebuilding operation.[7] These materials—the vast majority of which were never meant for public distribution and remain unarchived—do not fit easily with benign accounts of US militarism in the early days of the Iraq War. Although American spokesmen characterized the zoo's eventual rehabilitation as an unqualified success, the Baghdad Zoo faced a steady stream of challenges, from a lack of food and medicine to the routine threat of accidental bombardment. Most precarious of all was the fate of the animals themselves, whose experiences of violence and privation threatened to undercut American claims of martial benevolence.

For this reason, my approach in what follows is to underscore the corporeality of Baghdad's zoo animals: their messy biology, their inconvenient livingness, their hunger and thirst and pain. Baghdad's zoo animals, like those seeking to help them, existed as embodied subjects, whose bodily needs (for food, shelter, water, and medicine) played an outsize role in shaping the direction of the rebuilding effort. Even as the war raged, Baghdad's surviving zoo animals ate and drank, defecated in their cages, lolled in the heat, and sought refuge in the shade. They exuded odor, barked and growled, and sniffed the hot desert air. They took up space, whether alive or dead. And as with the population of "liberated" Iraqi citizens, their post-invasion behavior did not necessarily align with the optimistic predictions of the war's chief advocates.

Ultimately, this chapter offers yet another example of the mutual but unequal vulnerabilities of people and animals in the modern world. Well before 2003–4, the fates of Baghdad's human and zoo animal populations were deeply intertwined—by international sanctions, pollution, blistering heat, inadequate health resources, and authoritarian domestic politics. With the US invasion in 2003, Iraq's human and zoo animal populations faced new challenges—above all else, an inescapable susceptibility to physical and mental trauma. The main difference is that the animals in the zoo could not fight back (at least, not for long).

THE BAGHDAD ZOO: A BRIEF ANTHROPOCENTRIC HISTORY

At first glance, few institutions are more closely tied to the values of the Anthropocene than the modern zoo. Since their birth in the late eighteenth century, zoos have reflected all of the practices and modes of thinking—domestication, commodification, colonialism, nation-state competition, and so forth—we associate with human-driven environmental change.[8] Over the following century, zoos served as temples of European imperialism, living storehouses of an exotic nature (and exotic peoples) now harnessed to Western control. Writing about Victorian-era British zoos, historian Harriet Ritvo observes, "The maintenance and study of captive wild animals, simultaneous emblems of human mastery over the natural world and of English dominion over remote territories, offered an especially vivid rhetorical means of reenacting and extending the work of empire."[9] Starting in the early 1900s, some of the world's leading zoo architects turned to "barless" exhibits and other visual fakery to convey a vision of wild nature untouched by human activity.[10] Today, in a world of rapid climate change and species collapse, zoos tend to promote themselves as buttresses against animal extinction—human-constructed solutions to human-produced environmental destruction.[11]

Nevertheless, even contemporary zoos cannot fully disguise their Anthropocentric roots (nor would their advocates hope to do so). All zoos, no matter how seemingly "humane," remain wedded to ideas about power: specifically, the power of humans to raise, cage, display, reproduce, and dispose of animals as they see fit. Throughout the nineteenth century, many European zoos resembled little more than open-air prisons—clusters of cages, cells, and outdoor "habitats" designed to convey a symbolic and literal human-animal divide. (Well into the twentieth century, a number of European and North American zoos also featured displays of Indigenous peoples from Africa, South Asia, and the Arctic, fueling myths of native "savagery" and, in the process, working to justify Western ideas about the necessity of imperial power.)[12] To this day, zoos control nearly every aspect of animals' existence: what they eat, where they sleep, how often they breed, when and whether their lives have become expendable. Indeed, outside of medical laboratories and industrialized meat factories, it is difficult to imagine a more naked example of humans' perceived supremacy than a city zoo.[13]

Ultimately, zoos' most powerful function in the Anthropocene is to naturalize human dominion over the animal world.[14] Although contemporary zoos often frame their mission in terms of animal conservation or "wildlife heritage," zoos inevitably put people first—their safety, their needs and desires, their pleasures. Zoos are human institutions, beholden to economic forces, entangled in webs of politics and social change. This is most evident in times of war or national trauma, when zoos frequently alter their best practices and institutional missions to align with contemporary pressures. Wartime zoos, including the Regent's Park Zoo in London at the start of World War II, have been known to kill off large numbers of their animal stock to appease public hysteria about animal escapes.[15] Yet, even in moments of social stability, the modern zoo relegates nonhumans to a secondary status. Critic John Berger makes this point in his influential 1977 essay "Why Look at Animals?": "However you look at these animals, even if the animal is up against the bars, less than a foot from you, looking outwards in the public direction, *you are looking at something that has been rendered absolutely marginal*; and all the concentration you can muster will never be enough to centralise it."[16] Hence the irony of all zoos: the very act of caging animals now makes it easier to justify similar acts in the future.

The Baghdad Zoo was not immune to such dynamics. When it opened in the early 1970s, a time of relative prosperity in Iraq, the Baghdad Zoo was meant to serve as an object of civic pride, evidence of Iraq's emergent status as a modern power in the region. Although menageries date back millennia, most Iraqis would have never visited a zoo of the sort common in the West, let alone one in their nation's capital.[17] In the decades following its launch, the Baghdad Zoo would eventually attract hundreds of thousands of visitors a year (few other places in the city featured such an "abundance of trees, vegetation, and exotic animals," according to one military observer).[18]

From the beginning, however, the zoo struggled to meet the nutritional, behavioral, and psychological needs of the animals in its care. Many enclosures consisted of little more than dusty metal cages, and most of the zoo's seven hundred or so captive inhabitants were forced to spend their days baking in the hundred-degree heat. When American troops first entered the park, the zoo's pair of male tigers lived in a 230-square-meter "traditional Victorian type" enclosure surrounded on three sides with metal bars. The bottom of the enclosure was a mixture of sand, scrub grass, and bare dirt. For enrichment, the animals had access to a few small logs for scratching (added after the invasion), some scattered rocks, and a rectangular concrete pool, perhaps a foot deep and filled with soupy green water.[19]

Such levels of seeming indifference were by no means unique to zoos in Iraq (or the Middle East, for that matter). In 1984, for example, Zoo Atlanta lost its accreditation following high-profile accounts of animal cruelty and malpractice. Investigations revealed stories of unlicensed veterinary care, monkeys driven neurotic by neglect, and "hygiene so poor that a white-tailed deer was found 'struggling to walk' through mud and feces in its pen."[20] Unlike zoos in wealthier nations, the Baghdad Zoo also suffered the combined effects of war (against Iran and, in the early 1990s, a US-led coalition) and economic sanctions, which wrecked the Iraqi economy and made it difficult for Baghdad Zoo staff to provide adequate supplies of food and medicine. By the 1990s, however, Western media came to view the zoo's wretched condition as living proof of Saddam Hussein's brand of decadent cruelty. An article from 1995 called the zoo a "prison for animals"; another, an "animal slum, a metaphor for this once-powerful and predatory nation."[21] This image paved the way for what would become the dominant narrative of 2003–4: a rescue story with animal-loving Americans in the prime role.

RESCUING BAGHDAD'S ZOO ANIMALS

That narrative began in early April 2003, during what would later become known as the Battle of Baghdad.[22] Located in a public park not far from the presidential palace, the zoo had few natural defenses. Troops loyal to Saddam took up positions just outside the zoo gates, leaving animals unfed and, for the most part, helpless against the encroaching violence. By the time American troops finally secured the area, after more than a week of fighting, the zoo grounds were in shambles. Of the zoo's roughly 650 inhabitants, only a few dozen remained alive. At one point, a small pride of lions managed to escape the zoo grounds only to be gunned down by US Marines. Other zoo denizens were trapped beneath rubble or were suffering the effects of dehydration, heat exhaustion, and post-traumatic stress disorder (PTSD).[23]

At this point, the rescue story shifted to stage 2: bringing order to the chaos. Working with Dr. Adel Salman Mousa, the zoo director, and Lawrence Anthony, a South African conservationist who headed the rebuilding effort, American troops rounded up loose animals and secured the zoo grounds. They also crafted a makeshift holding pen of razor wire—known as the Baghdad Zoo jail—to detain those seeking to ransack the collection. From the very beginning, many volunteers singled out looters as the primary threat to the animals' safety. Lawrence Anthony blamed the destruction on "deranged mobs, hell-bent on ripping their city to shreds in a savage, greed-fueled spree," a characterization that would continue to inform Western attitudes throughout the zoo-rebuilding efforts.[24] Nevertheless, the most pressing concern was keeping the animals alive. That meant, above all, making sure they had access to (reasonably) potable water, without which some animals would die within a few days.

Not long after, the zoo occupation entered the most important stage of all: restoration and rehabilitation. Even as fighting still raged elsewhere in the city, a coalition of the willing, including NGOs from Britain, South Africa, and elsewhere, began to rebuild. They set up new generators and water filtration systems, provided food and much-needed medical treatment, and lectured local zoo workers on modern techniques of animal breeding and husbandry. In May 2003, a delegation of US military personnel, foreign aid workers, and Baghdad Zoo staff visited Luna Park, a private amusement facility about an hour north of the Green Zone. There, after paying off the park's owner, they retrieved some sixty caged animals—malnourished, diseased, many with obvious signs of self-mutilation—and brought them back to Baghdad, where they joined the rest of the collection.[25]

The zoo's most famous residents came courtesy of Uday Hussein, at the time attempting to avoid capture by US forces. For years, the dictator's son had maintained a private menagerie stocked with big cats—icons of wealth to be showcased alongside the rest of his treasures. Like other would-be strong men (Venezuelan president Juan Vincente Gómez Chacón, Ukrainian president Victor Yanukovych, Gambian president Yahya Jammeh, even Columbian drug kingpin Pablo Escobar), Uday and his father flaunted their animal collections as symbols of their ferocity and power. Explained one zoo employee, Uday "was a sick man, and he kept lions and tigers just to show his manhood, to show everyone that he cared more about animals than people. But he amputated their claws, and he took away their freedom, just like the people." This quote, promoted by the White House on its *Liberation Update* website, echoed a common refrain from the early days of the occupation: that the Hussein regime had treated people like animals (caged, mutilated, and degraded) and animals like personal playthings.[26] American troops confiscated Uday's cats and, in a metaphor for the broader US mission, released them into the public zoo. American media sources joyously recycled photos of American troops, standard-bearers of bemused innocence, standing outside the iron cages gazing at the dictator's son's cheetahs. In all, the United States spent some $2.15 million on the zoo effort, much of it aimed at raising the enclosures to internationally recognized standards of size and enrichment.[27]

Why all the bother? Why, in a time of such chaos, would the United States—indeed, anyone—devote resources to a zoo that, by all measures, should have been demolished years ago? For some, it was a matter of principle. "This was more than a zoo in a war zone," South African conservationist Lawrence Anthony wrote in 2011. "It was about making an intrinsically ethical and moral statement, saying: Enough is enough. You just can't say to hell with the consequences to the animal kingdom."[28] According to Wassem Sarih Ameen, a veterinarian at the zoo, many zoo staff saw their work through a lens of national reclamation, taking "particular pleasure" in their efforts to exhibit the stallions and exotic cats once owned by Saddam's family.[29] Early volunteers also felt genuine sympathy for the creatures they encountered. "Here they are, caged up, no water or food, no way to get water or food," one US officer explained to CNN in April 2003. "And, I mean, I just don't think it was something we turn our back on."[30] To do nothing while Baghdad's zoo animals died, hungry and bug-ridden and traumatized, was not an option.

At the same time, the zoo-building scheme functioned as a form of what Alison Howell and Andrew W. Neal call "imperial tutelage," a project of education and uplift aimed at "bring[ing] a particular kind of 'humane' liberal order to bear upon the Iraqi population."[31] This line of thinking suffused the early (optimistic) days of the invasion, when so-called humanitarians with guns sought to combat Saddam's hold on Iraqis with an onslaught of Western-style cultural preservation efforts. At the zoo, visitors would learn Western values about animal welfare and the "proper" relationship between humans and the nonhuman world.[32] Above all, advocates hoped that a revitalized zoo would serve a civilizing function in the region—a kind of counterinsurgency of the soul that would undo decades of state-sanctioned repression and ignorance.

The case of Sa'dia, a thirty-two-year-old brown bear housed at the Baghdad Zoo, provides a useful illustration of the kind of compassionate, technologically rooted hegemony Americans hoped to bring to Iraq. One of the few large animals to survive the early days of the US invasion, Sa'dia suffered from a cancerous tumor lodged in her abdomen. When it came time to remove the tumor, a team of American veterinary technicians and Iraqi zoo staff gathered inside the bear's cell. After aestheticizing Sa'dia with a dart gun, they shaved the animal's fur, cut away the tumor, and sutured the incision. All the while, both Americans and Iraqis documented the event for posterity (fig. 8.3).

American propagandists hoped that Sa'dia's surgery and similar acts might alter Iraqi impressions about US intentions in the region. Discussions of US-Iraqi zoo cooperation frequently mirrored the "when they stand up, we'll stand down" sloganeering of military officials. "The veterinarians in Iraq are victims of professional isolation," explained Lt. Col. Jose Lozada, a veterinarian with the 352nd Civil Affairs Command, at the time of the surgery. "Our hope is that, by involving zoo staff members and veterinarians in surgical procedures and vaccination processes, they quickly will regain control over this invaluable facility and its inhabitants."[33] In this scenario, Americans would help the Iraqis help themselves, and animals like Sa'dia would benefit in the process.

Animal-related imperial tutelage was not limited to the Baghdad Zoo. In February 2004, local veterinarians, alongside US soldiers and with funding from the military, launched the Iraqi Society for Animal Welfare (ISAW). Modeled on the Humane Society in the United States, the organization was devoted to the topic of "animal control"—specifically, tackling the twin problems of canine overpopulation and disease. The society hoped to change Iraqi thinking about dogs through its adoption and spay-and-neuter campaigns. Heading the new center was Dr. Farah Murrani, a young Iraqi veterinarian who simultaneously served as the zoo's associate director. Unlike many of her follow Iraqis, Murrani strongly supported the US invasion (at least at first), viewing it as an opportunity to revise local attitudes toward animal welfare. Equally notable was her willingness to break cultural taboos about touching dogs. (Dogs are considered "unclean" in some parts of the Muslim world.) With Murrani at the helm, the ISAW represented more than an effort to cut down on strays. It was, according to a military spokesman, a "first step toward a safer and animal-friendly country."[34]

Above all else, the zoo-rebuilding campaign aligned with Bush administration–era ideas about American empire-through-example. In the eyes of its supporters, the United States was a humane hegemon—the kind that will blow up a house on one day and pay the residents for the damage the next.[35]

FIG. 8.3. One of many photographs taken at the scene, this image documents a US-led surgery to remove a cancerous tumor from the abdomen of Sa'dia, a thirty-two-year-old brown bear at the Baghdad Zoo. Among those pictured are US Army colonel Mark Gants (bottom right in medical smock), SPC Erin McLoughlin (bottom left, kneeling), and Dr. Adel Mousa (top, second from left), the director of the Baghdad Zoo. While individual actors were undoubtedly sincere in their efforts to help Baghdad's zoo animals, such images contributed to the broader US propaganda campaign to win over Iraqi sentiment. Moreover, these specific photos would have lent credence to the "surgical" metaphors (i.e., healing through controlled bloodletting) that proliferated during the war's early days. Records of the Office of the Secretary of Defense, 1921–2008, National Archives.

Americans hoped to transform the values of Iraqi society and install, alongside a belief in democracy, a host of ideas about how society should be ordered. In this sense, as Alison Howell and Andrew W. Neal observe, "rebuilding the zoo was not only about winning hearts and minds in the shallow sense of assuring the population's acquiescence, but about reforming and liberalizing their very hearts and minds."[36] Put bluntly, the military's zoo work was part of a larger ideological mission, as necessary to US counterinsurgency efforts as kicking in the doors of suspected terrorists.

This effort reached its fruition—at least as far as the media was concerned—on July 19, 2003 (fig. 8.4). On a cloudless day, with machine gun–wielding soldiers manning the front gates and a thin ribbon of balloons floating overhead, the zoo was officially reopened to great applause. In grainy video footage taken on that day, Farah Murrani pans over a scene of children dancing and

FIG. 8.4. Taken on July 19, 2003, this photograph captures the mix of excitement and tension that pervaded the zoo's reopening. As music blared from speakers, men, women, and families strolled the palm-lined lanes, inspecting the rows of caged animals. All the while, uniformed US troops armed with assault rifles patrolled. Image courtesy of William Sumner.

leaping about to the tempo of beating drums. "This is amazing, I swear," she narrates.[37] Lawrence Anthony viewed the day as the "first tentative step toward normality." One US official at the time declared, "The park's reopening is symbolic of the new freedom in Iraq. The zoo is a place for the community to go and reflect on the meaning of life."[38]

And right about here is when the media narrative (the rescue story) begins to peter out. Journalists and other observers would check in from time to time—at the death of the tiger, for example—but the contours of the story were essentially set: once a metonymy of an "invaded nation," the Bagdad Zoo was now a shining example of what liberal interventionism could bring.

BEHIND THE BARS

The reality, however, was more problematic, not least because it proved impossible to separate the politics of the war—the propaganda needs of the US campaign, the internal divisions of Iraqis, America's profound ignorance of the region, and so forth—from the efforts on the zoo's behalf. A general atmosphere of suspicion pervaded all security matters, and some Americans viewed their Iraqi collaborators with unease. More than six months after she started working at the zoo, associate director Farah Murrani, one of the primary liaisons between Iraqi staff and US military forces on the ground, still had to wait in line outside the Green Zone for lengthy security checks. At one point, the delays were so long that they threatened the lives of a group of nine lions temporarily housed at a remote site near the brigade headquarters. The military officer in charge, William Sumner, had to personally accompany her to the front of the line, but even then, she was denied the level 2 clearance that would have allowed her to carry out her job.[39]

Complicating matters further, the military was not nearly as supportive as both they and journalists would later claim. Throughout April 2003, the effort to rescue Baghdad's zoo animals lacked any clear sense of mission—beyond trying to keep more animals from dying. Much of the life-and-death work of protecting the zoo's animals was carried out by a relatively small group of civil affairs officers, off-duty soldiers, and Iraqi volunteers. Higher-ups in the CPA eventually came on board once they saw the propaganda value of the effort. From the very beginning, however, matters of animal safety and health often took a backseat to the goal of declaring mission accomplished and reopening to the public.

Moreover, it was never quite clear what kind of zoo they were aiming to create. Was the goal to build a zoo on par with first-tier zoological institutions around the globe? Or should they aim for something more modest? Politics and practicality dictated that the Baghdad Zoo would have to play multiple, sometimes contradictory, roles at once: a last-ditch refuge for abused animals, a playground for the capital's war-weary families, an emblem of national renewal, and a showcase of American benevolence. Given the circumstances, Brendan Whittington-Jones, a volunteer from South Africa's Thula Thula Private Game Reserve, argued that it made little sense to try to ape European or American zoos. Because of its lack of resources, Baghdad was better off sticking to "desert species," the kinds that can "generally be found in captive facilities throughout the Middle East." Although Iraq's indigenous animals were not especially "charismatic or aesthetically beautiful," the South African nonetheless hoped they might generate local interest in "Iraq's severely damaged natural heritage." He even advocated for the zoo to accept locally owned "problem animals," although he drew the line when it came to the black market. Still, Whittington-Jones worried (rightly, it turns out) that such measures would satisfy neither zoo-goers nor zoo reformers. With time, they would insist upon "exotic species" typically associated with twentieth-century Western zoos.[40]

Barbara Maas, of the UK-based nonprofit Care for the Wild International, echoed these concerns. Arriving in Baghdad a few weeks after the start of the invasion, she was horrified by the scenes of animal suffering. She petitioned the authorities to allow her to remove one of the bears—its eyes blinded by cataracts, its days spent pacing back and forth in a small concrete cage—to a sanctuary in Greece, already home to thirteen rescued dancing bears. Maas urged the zoo to abandon keeping bears on "compassionate and humanitarian grounds," arguing that many Western zoos had already started "phasing out" ursines from their collections. She also suggested sending at least six of the zoo's nineteen lions—a high number even for large zoos—to a sanctuary in South Africa. (Emirates Airlines had agreed to fly them out at a discount.)[41] However, zoo management refused, and Maas was "asked not to return" to Baghdad.[42] In the battle between animal safety and zoo rebuilding, zoo rebuilding won.

On the whole, people involved in the rebuilding effort tended to sidestep the messier ethical considerations of animal captivity, focusing instead on the more pragmatic details of keeping the zoo up and running (e.g., feeding animals, maintaining enclosures, paying and training staff). They also carried out a rearguard battle against those outside Iraq who, purposefully or not, threatened to undermine their mission. At one point, William Sumner went so far as to write the Doris Day Animal Foundation (DDAF), asking it to replace a photo on the group's website. It seems that the DDAF had mistakenly posted an image of one of the zoo's cheetahs under the heading "Success Story: The U.S. Senate Passes a Bill to Protect Wild Cats from Ending Up in the Exotic Pet Trade." According to Sumner, "Every good piece of work we do seems to be followed by some calamity," and associating the Baghdad Zoo with the black market certainly did not help.[43]

Complicating the zoo restoration efforts were the animals themselves—flesh-and-blood, breathing, eating, stinking beings whose needs and desires, illnesses and anxieties, behavior and regard for visitors did not necessarily match up with the ambitions of their keepers. As zoo workers learned in Baghdad (and elsewhere in Iraq), many zoo animals have an all-too-stubborn ability to survive in the most abject of conditions. One of the great tragedies of animal captivity is the fact that a number of animals do not *need* fresh air, clean water, and living space to live painfully long lives. In Luna Park, the coalition of volunteers discovered a malnourished gazelle resting atop a pile of garbage and a pair of porcupines that had not been fed for days.[44] Throughout the first year, many of the veterinarians' surgical interventions were meant to address long-standing ailments that, until now, Baghdad's zoo animals had been forced to endure. That said, death was a regular feature at the Baghdad Zoo, despite the reformers' best efforts. Between July and mid-October 2003, the zoo suffered eleven major animal deaths, including those of three young lions (two from pneumonia, one from cancer), a brown bear (septicemia), a goat (listeriosis following a bacterial infection), a wild boar and a hawk (both wounded in fights), and of course, the tiger fatally shot by a US soldier. Of the eleven dead, only one—a pony—expired because of what zoo staff deemed "old age."[45] All of these animals would need to be removed from their cages, buried or burned, and ultimately replaced.

ZOO CRIME II: KILLING A PUPPY

In retrospect, one aspect of the Baghdad Zoo in 2003 should not be surprising: the violence against the zoo animals. All zoos have to grapple with violence of one sort or another. Indeed, perhaps the most persistent threat to zoo animal safety is the casual, frequently "joking" abuse from zoo visitors. Zoo directors have long attempted to ascertain the root cause of animal abuse.[46] (Some have described it as a kind of cultural sickness, a disease that infects modern society.)[47] Moreover, the nature of zoos seems to encourage animal abuse, despite keepers' intentions. The very act of keeping animals caged signals their difference, their lesser-than status. They exist primarily to be gawked at, exposed, and presented for our entertainment; within this environment, the line between "looking" and "injuring" is increasingly blurred. Added to this situation is a hierarchy among the animals themselves—a ranking in which some are valuable, some are disposable; some eat, others are eaten.

This permission structure is amplified during wartime, when many of the rules that govern civil society—about property rights, the use of force, individual sovereignty—are already weakened, if not eliminated altogether. On January 5, 2004, less than four months after the tiger incident, this reality hit home at the Baghdad Zoo.

On that day, around 1:30 in the afternoon, a group of soldiers in Humvees roared into the zoo. Emerging from their vehicles, they caught a five-month-old puppy that lived on the zoo grounds, zip-tied its muzzle and hind legs, and pushed it through the bars of the lion enclosure. The puppy dropped about ten feet into a concrete moat, at which point the lions attacked (fig. 8.5). In a matter of seconds, the soldiers had turned the puppy from a pet into an object of sadistic amusement, beyond the limits of ethical consideration. When confronted by a zoo vet, they claimed to be "just having some fun."[48]

Upon learning of the soldiers' actions, Farah Murrani immediately emailed seventeen local allies and NGOs that had contributed to the zoo's rehabilitation. "I ask you to help me to get this poor animal's rights from those moron merciless soldiers," she wrote. "Do they think just because they are in Iraq? What will happen to me if I [throw] one of the soldiers to the lions just for having fun?" Murrani raged against Americans' seeming indifference to the damage they left in their wake: "The tiger incident has been forgotten by everybody & we didn't even get a formal apology from any one in charge." Most infuriating

FIG. 8.5. This image shows the immediate aftermath of the puppy episode. In the center of the photo, we can see the body of the puppy (now dead). It seems that zoo staff were less upset by the fact that the lions consumed a live animal than the fact that they consumed *this* live animal, one that was viewed as a pet/companion rather than as soon-to-be food. Image courtesy of William Sumner.

of all was the smug hypocrisy that seemed to run through all aspects of the US mission in Iraq: "From one side I'm helping American families to adopt dogs and sending them home for them and on the other side they are killing my pets and people's pets. . . . I will not stop until I get revenge for the puppy, until I kick somebody's guts out for doing this." Murrani signed off (to her readers but in some respects to all Iraqis) with an ominous warning: "If you saw a soldier around you walk away, you may be the next meal for the lions."[49]

It is worth pausing here to consider several points. One is the parallel between the behavior of these soldiers and that of soldiers stationed at another carceral institution, men and women who also took pleasure from (and took pictures of) the abuse of those perceived as disposable, beyond human empathy. In the nearly two decades since its revelation, the systemic abuse in 2003 of prisoners at Abu Ghraib prison, about twenty miles west of the capital, remains a low-water mark in the US conquest of Iraq. The publication of photographs of torture, sexual sadism, and gross violation of human rights undermined all the noble rhetoric used to justify the US presence in the region. In the eyes of the torturers, Iraqi prisoners belonged to a class outside ethical consideration—disposable, unhuman—a point driven home in a photo of a naked detainee posed with a dog chain around his neck.[50] In the weeks following the disclosure, the US military disciplined the "bad apples" responsible for the torture, doubling down on the message of humane interventionism that had dominated much of the propaganda campaign.[51] Still, as with the puppy killing, Abu Ghraib seemed to reveal a dark undercurrent in American war culture: that (some) US troops enjoyed killing; they got their kicks from torturing the most precarious, human and nonhuman alike.[52]

In addition, Murrani's anger at the puppy killers' "fun" inadvertently hit on the contradiction at the heart of all "good zoo" discourse. No zoo, no matter how well equipped or internationally recognized, can fully escape the violence inherent in its history and makeup (the violence of acquisition, the violence of captivity, the violence of animal "disposal," and so forth). While they did not condone throwing puppies into the lions' cages, leaders at the Baghdad Zoo justified doing the exact same thing to donkeys, sometimes live. For Lawrence Anthony, it was "a straight mathematical equation, conservation politics in the face of circumstance."[53] But live feedings also functioned as a kind of publicity stunt—animal terror as spectacle and entertainment. To this day, YouTube features numerous videos of live-animal feedings at the Baghdad Zoo and elsewhere, many complete with a soundtrack of cheering onlookers.[54]

CONCLUSIONS

Looking beneath the optimistic media narrative about the Baghdad Zoo in the first year of the US occupation, we can draw at least four broad conclusions. First, the rebuilding of the Baghdad Zoo was never as seamless, as ideologically driven, or as narratively coherent as it appears in Western media. Viewed historically, the story of the Baghdad Zoo, much like the Iraq War itself, appears successful only when viewed through a narrow lens of 2003–4. Expand that lens farther—say, to the present day—and you will find an underfunded, conflict-ridden zoo that suffers from many of the same problems as in Saddam's time.

Second, the story of the military's rebuilding efforts demonstrates the multiple levels upon which human-animal relations simultaneously play out. For those working on the ground, saving the zoo was an effort to preserve the lives of very specific animals—individual creatures whose pain and suffering were more than mere abstractions. Whatever one thinks about the Iraq War, it is impossible to dismiss the motivations of those risking their lives to carry out zoo rehabilitation work. And it was *work*—feeding, cleaning, tending, and building—all against the backdrop of a looming insurgency. At the same time, the military's efforts on behalf of the Baghdad Zoo must be seen as part of a larger effort to justify and legitimize the United States's military intervention. In rebuilding the zoo, occupying forces saw an opportunity not only to save (or, at least, lengthen) the lives of vulnerable animals but also to demonstrate the United States's reputation as a force for good in the region.[55]

Third, what happened in Baghdad in 2003–4 highlights the reluctance, even among certain "animal lovers," to think about animals beyond a framework of captivity. Faced with the abject conditions of the prewar zoo, as well as the dismal state of animal collections elsewhere in the country, volunteers and military officials did not shutter the zoo, as some activists had hoped. Nor did they attempt to limit the zoo's holdings of large cats and bears, despite the inherent difficulties associated with their care. If anything, the US effort focused on rebuilding the same type of facility that had existed prior to 2003—only (somewhat) more humane.

Most important of all, this brief history reminds us of the mutual, albeit unequal, vulnerabilities of occupied peoples and captive animals. In 2003–4, Baghdad's zoo animals found themselves at the mercy of a military force with the power of life and death, a regime that viewed their modified captivity as a sign of progress and that struggled to conceive of anything other than a cleaned-up version of what had already existed. Much like the detainees at Abu Ghraib, the animals at the Baghdad Zoo endured a precarious existence, in no small part because of the violence inherent in all carceral regimes. In this sense, the story of the Baghdad Zoo was just a chapter in a drama playing out across Iraq—a story that would end in half a million Iraqis dead, a society in shambles, and a small city menagerie still struggling to stay afloat.[56]

UPDATE: FIVE YEARS LATER

On August 4, 2008, a pair of Bengal tiger cubs arrived at the Baghdad Zoo—replacements for the one killed nearly five years earlier. They were gifts from a North Carolina animal sanctuary, and ambassador Ryan Crocker had spent $66,000 of taxpayer funds shipping them via DHL to Iraq. Adel Mousa, the zoo director, praised the military's efforts to restore the zoo "after years of degradation and misery." The zoo was, according to one account, now "approaching normality," a claim that would be repeated, off and on, for the following decade.[57] Not all outside observers shared such optimism, however. On learning of the shipment, Lisa Wathne, of the People for the Ethical Treatment of Animals (PETA), opined: "These tigers will be caged, helpless, and completely dependent on humans to survive in an area where many people live in fear and are still without access to basic necessities."[58] (The same might be said about the vast majority of captive animals around the world.)

And Keith Mitchell, the one-time "international poster boy for the misbehavior of U.S. troops in Iraq"?[59] Officially exonerated by a court martial, he was medically discharged in 2006 and died from "complications of diabetes" a year later, most likely as a result of the twenty-plus surgeries to restore his mangled hand.[60]

NOTES

1. This narrative of the tiger killing is derived from Department of the Army, "Death of a Tiger at Baghdad Zoo, Zawra Park," 354 Civil Affairs Brigade, September 20, 2003, author's collection.

2. David Armstrong, "Zoo Story: Sergeant Mauled by Iraq Tiger Gets Some Vindication," *Wall Street Journal*, November 18, 2003, A1.

3. In a recent article, researchers Céline Albert, Gloria M. Luque, and Franck Courchamp identify charismatic megafauna (or "flagship species") with six traits: "rare," "endangered," "beautiful," "cute," "impressive," and "dangerous." See "The Twenty Most Charismatic Species," *PLOS One* 13, no. 7 (2018), available at https://doi.org/10.1371/journal.pone.0199149.

4. Department of the Army, "Death of a Tiger at Baghdad Zoo."

5. Saddam Hussein Abd al-Majid al-Tikitri, known in the West as Saddam

Hussein (or simply Saddam), ruled as president of Iraq from 1979 until 2003. Often supported by allies in the West (including the United States), Hussein's regime was characterized by authoritarianism, violent repression, and the theft of state resources by the dictator's immediate family. Following the US-led invasion, Saddam Hussein was tried for crimes against humanity and executed by hanging in December 2006.

6. The Coalition Provisional Authority (CPA) was the US-led transitional caretaker force that governed Iraq for fourteen months after the 2003 invasion. Along with administering all government functions, the CPA was charged with "reconstructing" the nation's civil and physical infrastructure. On the numerous problems with the CPA, see Thomas E. Ricks, *Fiasco: The American Military Adventure in Iraq* (New York: Penguin, 2006).

7. For the sake of confidentiality, I have chosen to withhold the name of the sources of these materials. Nevertheless, all of them derive from sources within the military, from the CPA, or from civilians working on the rebuilding of the Baghdad Zoo in 2003–4.

8. See the introduction to this volume, "The Mule in the Coal Mine."

9. Harriet Ritvo, *The Animal Estate: The English and Other Creatures in the Victorian Age* (Cambridge, MA: Harvard University Press, 1987), 205.

10. The innovation of such exhibits, notes zoo historian Nigel Rothfels, is their ability to convey the narrative that "animals have been put in zoos increasingly because they are nice, healthy, safe places to be and because the animals, we are told, might be better off there than in the real world." See Nigel Rothfels, *Savages and Beasts: The Birth of the Modern Zoo* (Baltimore: Johns Hopkins University Press, 2002), 199.

11. On the history and future of zoos and animal conservation, see Ben A. Minteer, Jane Maienschein, and James P. Collins, eds., *The Ark and Beyond: The Evolution of Zoo and Aquarium Conservation* (Chicago: University of Chicago Press, 2018).

12. This topic is far too broad to discuss here except to say that zoos have always functioned as instruments of hierarchy, drawing distinctions between one group of living subjects (those who see, who have the power of free movement, who presumably require no physical confinement) and another (those who are seen, who do not have the power of free movement, and who require physical confinement). On the practice of displaying non-European peoples in "human zoos," see Pascal Blanchard, Nicolas Bancel, Eric Deroo, and Sandrine Lemaire, eds., *Human Zoos: Science and Spectacle in the Age of Empire* (Liverpool: Liverpool University Press, 2009); Rothfels, *Savages and Beasts*, 81–142.

13. Geographers Chris Philo and Chris Wilbert have characterized labs and breeding stations as part of a "hidden geography . . . wherein many animals live and die as part of a highly unequal human-animal relation predicated on the utility, adaptability and expendability of the animals so incarcerated." This description might be extended to most zoos, despite their purported publicness. Chris Philo and Chris Wilbert, "Animal Spaces, Beastly Places: An Introduction," in *Animal Spaces, Beastly Places: New Geographies of Human-Animal Relations*, ed. Chris Philo and Chris Wilbert (London: Routledge, 2000), 2.

14. On the concept of dominion and animal cruelty, see Matthew Scully, *Dominion: The Power of Man, the Suffering of Animals, and the Call to Mercy* (New York: St. Martin's Press, 2002).

15. See John M. Kinder, "Zoo Animals and Modern War: Captive Casualties, Patriotic Citizens, and Good Soldiers," in *Animals and War: Studies of Europe and North America*, ed. Ryan Hediger (Leiden: Brill, 2013), 45–75.

16. John Berger, "Why Look at Animals?," in *About Looking* (New York: Vintage, 2001), 24.

17. The founding of the Baghdad Zoo did not translate into widespread zoo attendance across the country. A 2003 questionnaire distributed to Baghdad Zoo visitors found that fewer than 50 percent had ever been to a zoo in the past. Brendan Whittington-Jones, "Baghdad Zoo Status Report," December 1, 2003, author's collection.

18. Whittington-Jones, "Baghdad Zoo Status Report."

19. I have drawn my description of the tiger enclosure from photographs taken by US personnel and from Whittington-Jones, "Baghdad Zoo Status Report." Although some critics believe that no zoo enclosure can fully meet an adult tiger's territorial, behavioral, and psychological needs, the Association of Zoos and Aquariums has published a 121-page manual of instructions for caring for tigers in captivity. See Association of Zoos and Aquariums, *AZA Tiger Species Survival Plan, Tiger Care Manual* (Silver Spring, MD: Association of Zoos and Aquariums, 2016), https://assets.speakcdn.com/assets/2332/tiger_care_manual_2016.pdf.

20. Art Harris, "Atlanta's Sorry Zoo," *Washington Post*, June 8, 1984.

21. Quoted in Alison Howell and Andrew W. Neal, "Human Interest and Human Governance in Iraq: Humanitarian War and the Baghdad Zoo," *Journal of Intervention and Statebuilding* 6, no. 2 (2012): 218, 219.

22. This general narrative has been reprinted in many sources, notably Lawrence Anthony with Graham Spence, *Babylon's Ark: The Incredible Wartime Rescue of the Baghdad Zoo* (New York: Thomas Dunne Books, 2007); and Kelly Milner Halls and Major William Sumner, *Saving the Baghdad Zoo: A True Story of Hope and*

Heroes (New York: Greenwillow Books, 2010). On post-traumatic stress disorder (PTSD) in animals, see Cari Romm, "Animals Can Get PTSD, Too," *The Cut*, May 5, 2016, https://www.thecut.com/2016/05/animals-can-get-ptsd-too.html; Ruchi Kumar, "Can Animals Suffer from PTSD?," *Washington Post*, July 8, 2016.

23. Interestingly, there is a structural similarity between early media reports about the zoo and those that accompanied the US liberation of concentration camps during World War II. Both described a two-part process that would be common in twentieth-century atrocity journalism: first, the moment of discovery, followed by denouncements of the inhumane cruelty of those who could perpetuate such horrors. On media accounts of concentration camp liberations, see John C. McManus, *Hell Before Their Very Eyes: American Soldiers Liberate Concentration Camps, April 1945* (Baltimore: Johns Hopkins University Press, 2015).

24. Quoted in Howell and Neal, "Human Interest and Human Governance in Iraq," 222.

25. Halls and Sumner, *Saving the Baghdad Zoo*, 14. The animal collection at Luna Park shocked and disgusted US military personnel. They were outraged not only by the horrific conditions endured by the animals but also by the fact that the zoo housed a number of animals that Westerners traditionally view as "pets," including several small dogs.

26. Originally found in a July 2003 story in the *Washington Post*, zoo employee Alaa Karim's quote was featured on the White House's Liberation Update site, a collection of news releases and tidbits aimed at putting a positive spin on the US war effort (https://georgewbush-whitehouse.archives.gov/news/releases/2003/10/20031022–9.html).

27. "Baghdad Zoo Answer to Monkey Surplus," *Tulsa World*, February 26, 2009, A6.

28. Anthony, *Babylon's Ark*, 50.

29. "Baghdad Zoo—Respite from the Urban Jungle," confidential cable dated February 21, 2008, WikiLeaks: Public Library of US Diplomacy, https://wikileaks.org/plusd/cables/08BAGHDAD501_a.html.

30. Quoted in "Baghdad Zoo: A Different Battle," CNN.com, April 17, 2003, https://www.cnn.com/2003/WORLD/meast/04/16/sprj.nilaw.baghdad.zoo/.

31. Howell and Neal, "Human Interest and Human Governance in Iraq," 224.

32. Brendan Whittington-Jones, a volunteer from South Africa's Thula Thula Private Game Reserve, made this point explicit in a December 2003 report on the zoo-rebuilding effort, which featured an epigraph often attributed to Gandhi: "The greatness of a nation and its moral progress can be judged by the way it treats its

animals." Baghdad Zoo Status Report by Brendan Whittington-Jones of the Thula Thula Private Game Reserve, Zululand, December 1, 2003, author's collection.

33. Update on Operation Iraqi Freedom, from *Army Reserve Magazine* 49, no. 4 (n.d.), author's collection.

34. Chad D. Wilkerson, "Two-Legged Help Stands Up Baghdad Welfare Facility," *DOD News*, February 4, 2004.

35. Meant as an expression of sympathy (rather than an admission of wrongdoing), the United States has distributed more than $50 million in monetary payments—variously known as sympathy payments, condolence payments, or *ex gratia* payments—to victims of US military action. According to Thomas Gregory, these payments are not a "tool of transparency or accountability" but a "weapon of war, something that can be used by the military to help manage the consequences of civilian harm and ensure that they do not disadvantage themselves in the battle for hearts and minds." Thomas Gregory, "The Costs of War: Condolence Payments and the Politics of Killing Civilians," *Review of International Studies* 46, no. 1 (2020): 159.

36. Howell and Neal, "Human Interest and Human Governance in Iraq," 225.

37. Farrah Murrani, film footage of zoo reopening, July 19, 2003, author's collection.

38. Anthony, *Babylon's Ark*, 231.

39. William Sumner, Memorandum for LTC Gainey, 354th Civil Affairs Brigade, Baghdad Iraq, January 3, 2004, author's collection.

40. When asked what kinds of animals they hoped to see, respondents named lions, elephants, dolphins, tigers, and bears as the most popular species. Whittington-Jones, "Baghdad Zoo Status Report."

41. Letter from Barbara Maas to the Baghdad Zoo, ca. 2003, author's collection.

42. Whittington-Jones, "Baghdad Zoo Status Report."

43. William Sumner to Doris Day Animal Foundation, ca. 2003–4, author's collection.

44. Mission Report: Luna Park Zoo, May 11, 2003, author's collection.

45. Zoo Death List, ca. late 2003, author's collection.

46. For a broad discussion of this topic, see Gareth Davey, "Visitor Behavior in Zoos: A Review," *Anthrozoös* 19, no. 2 (2006): 143–57.

47. Writing on the eve of World War II, one Wisconsin zoo director chalked up the sadism of "moronic visitors" to violent popular culture. "The zoo, like the school, has to combat the monkey in man and the false standards of conduct implanted into young minds by fiction writers, movie scenarios, stage heroes and crudely glorified historical characters," he railed. "So we have all sorts of young

Buffalo Bills, Kit Carsons, . . . Dillingers and Capones stalking the animals in the zoo and putting on airs of injured innocence when caught tormenting and hurting them." Ernest Untermann, "Modern Zoo Problems," Papers, 1933–1956, Wisconsin Historical Society, Milwaukee.

48. Memo for Col. Mitchell, Governance Team Chief, 354th Civil Affairs Brigade, Baghdad, Iraq by William E. Sumner, January 7, 2004, author's collection.

49. Email from Dr. Farah Murrani, Associate Director of the Baghdad Zoo, to NGOs, attached to Memo for Col. Mitchell, author's collection.

50. Andrew J. Bacevich, *America's War for the Greater Middle East: A Military History* (New York: Random House, 2016), 264–66.

51. On the torture of prisoners at Abu Ghraib, see Philip Gourevitch and Errol Morris, *The Ballad of Abu Ghraib* (New York: Penguin, 2009).

52. The dehumanizing treatment of Iraqi captives was not limited to Abu Ghraib. Qais Mohammad al Saliman, a fifty-four-year-old engineer, was forced to spend more than a month at an outdoor detention facility after he was wrongly arrested. Describing the conditions to a US journalist, al Saliman recalled, "It was very dirty and there were dust storms. There was no chance to wash before praying. There was only a hole for a toilet, in front of over 100 people. . . . The Americans came here talking about cooperation, and I was treated like an animal in a zoo in my own country." Dana Hull, "Iraqi Detainees Complain about Treatment," Knight Ridder, July 10, 2003, archived in Research Guide to the U.S. War on Iraq, https://www.robincmiller.com/iraq-rg.htm.

53. Anthony, *Babylon's Ark*, 63.

54. One such clip from the Baghdad Zoo, titled "Lions Eat Live Donkeys," has been screened nearly 400,000 times since 2016; see https://www.youtube.com/watch?v=S6HCdop5jCc. Importantly, a number of these videos feature Asian zoos, fueling preexisting stereotypes about the peculiar cruelty of animal treatment in non-Western nations.

55. Though he did not specifically address the zoo, President George W. Bush framed the US reconstruction effort in distinctly moral terms during a speech at Fort Bragg, North Carolina, in June 2005: "Rebuilding a country after three decades of tyranny is hard and rebuilding while at war is even harder. . . . [But] together with our allies, we will help the new Iraqi government deliver a better life for its citizens." "Full Text: George Bush's Iraq Speech," *The Guardian*, June 28, 2005.

56. A 2018 analysis in the *Washington Post* estimated more than 600,000 civilian casualties in the Iraq War. Philip Bump, "15 Years after the Iraq War Began, the Death Toll Is Still Murky," *Washington Post*, March 20, 2018.

57. Rob Nordland, "Tigers Return to Baghdad," *Newsweek*, July 7, 2008.

58. Quoted in Kim Gamel, "Rare Tigers Leap from N.C. to Baghdad," *Charlotte Observer*, August 9, 2008.

59. David Armstrong, "Zoo Story: Sergeant Mauled by Iraq Tiger Gets Some Vindication," *Wall Street Journal*, November 18, 2003, A1.

60. Nordland, "Tigers Return to Baghdad."

CONTRIBUTORS

VANESSA BATEMAN is a postdoctoral researcher at Maastricht University and received her PhD in art history, theory, and criticism with a specialization in anthropogeny from University of California San Diego. Her work analyzes the representation of animals in North American art and visual culture from the nineteenth century until the present.

JOSHUA ABRAM KERCSMAR is associate professor of environmental studies at Unity College. He has previously written on how dog evolution shaped Euro-Indian relations in North America. His current project explores how early American and Caribbean enslavers developed overlapping strategies for improving livestock and managing enslaved humans from 1600 to 1834.

JOHN M. KINDER is associate professor of history at Oklahoma State University, with a special interest in the effects of war on American culture and society. His current research focuses on how zoos—both in the United States and around the globe—have been transformed during periods of military conflict.

JENNIFER MARKS is a doctoral candidate in history at the University of Iowa and a technical writer for a transportation civil engineering firm in Portland, Oregon. Her forthcoming dissertation retells Chicago's rise as a metropolis at the turn of the nineteenth century through the city's human-animal relationships.

SUSAN NANCE is professor of history and affiliated faculty with the Campbell Centre for the Study of Animal Welfare at the University of Guelph in Ontario, Canada. Her research documents animals in US entertainment history; she is the author of various books, including *Rodeo: An Animal History* (2020).

ANDREA RINGER is an assistant professor of history at Tennessee State University. Her forthcoming book, *Circus World: Roustabouts, Lions, and Other Tented Workers, 1880–1980*, examines the circus as an interspecies workplace. More broadly, her research examines spaces of animal spectacle and has appeared in *Gender and History* and *Labor: Studies in Working-Class History*.

MARY TRACHSEL is associate professor of rhetoric at the University of Iowa, with an interest in human-nonhuman animal communication. Her research in animal studies is, in part, informed by her upbringing on an Iowa farm.

JESSICA WANG holds a joint appointment as professor of geography and professor of US history at the University of British Columbia, where she studies the intersections between science and state power. Her current research focuses on agriculture and the US insular empire in the early twentieth century.

INDEX

Page numbers in *italics* refer to figures.